LIFE COSTING

Theory and Practice

LIFE CYCLE COSTING
Theory and Practice

Roger Flanagan
George Norman
Justin Meadows
Graham Robinson

BSP PROFESSIONAL BOOKS

OXFORD LONDON EDINBURGH

BOSTON MELBOURNE

First published 1989

British Library
Cataloguing in Publication Data

Life cycle costing.
    1. Buildings.   Design.   Life cycle costing
    I. Flanagan, Roger
    690'.068'1

ISBN 0–632–02578–6

BSP Professional Books
A division of Blackwell Scientific
    Publications Ltd
Editorial Offices:
Osney Mead, Oxford OX2 0EL
    (Orders: Tel. 0865 240201)
8 John Street, London WC1N 2ES
23 Ainslie Place, Edinburgh EH3 6AJ
3 Cambridge Center, Suite 208, Cambridge
    MA 02142, USA
107 Barry Street, Carlton, Victoria 3053,
    Australia

Set by DMD Ltd, Oxford

Printed and bound in Great Britain

Contents

Preface

Progress in the application of life cycle costing techniques as an aid to decision-making in the building industry has been steady but not particularly inspiring. There is probably general acceptance now that these techniques will lead to better decisions on design – both of complete buildings and individual building components – on refurbishment, on maintenance planning and on management of the building portfolio. It has been shown in recent analyses of project planning discussions that life cycle costs do occupy an important part in the clients' initial thinking. Unfortunately, it also appears that, as the design process continues, the design team becomes increasingly preoccupied with initial capital costs, much of this preoccupation being client driven. The central message to be drawn from this kind of analysis is that considerable efforts are still required to convince clients and all members of the design team of the increased efficiency that will come from striking a proper balance between initial capital costs and future operating costs. There is no doubt that effective use of life cycle costing will confer considerable benefits on the clients of the industry, and so on the industry itself, but much remains to be done to remove the remaining barriers to such practical implementation.

We hope that this book will contribute to this process of moving life cycle costing from being a good idea to being a practical reality. Our intention has been to build on existing work in a number of ways with the intention of addressing three particular issues:

(i) the feeling that is often expressed that since life cycle costing deals with the future, and since the future is uncertain, the results of any life cycle costing calculation will inevitably be inaccurate;

(ii) problems of generating the data on which life cycle costing calculations are based;

(iii) the feeling that the seeming scientific basis of life cycle costing removes managerial discretion.

The most obvious point to make about the uncertainty issue is very simply that all decisions are based upon uncertain information, whether these decisions refer to estimates of initial costs or estimates of future operating costs. All that changes is the degree of uncertainty. More importantly, techniques now exist that allow us to *use* uncertainty in order to improve decisions. Just what this means we try to make clear in the text, first by providing an extended discussion of life cycle costing techniques that incorporate risk and uncertainty and secondly by showing how these techniques can be made to work in practical situations. An overriding priority of most clients is to avoid surprises: indeed, this will often be rather more important than searching for the lowest cost option. Risk analysis techniques can be used to identify the primary sources of potential surprises and to identify their likely impact. Once this kind of information is available, steps can be taken – perhaps by changing the design or the specification of components – to reduce potential uncertainties.

The 'data problem' is gradually becoming less severe. Many client and consultant organisations are now building up extensive and well designed data bases. Again, of course, much remains to be done but we are now confident that any organisation that wishes to apply life cycle costing to particular construction industry decisions will be able to generate the necessary data. What we try to do is show how such data bases can be constructed and also present an amount of 'live' data.

The idea that life cycle costing produces scientific answers that leave little scope for managerial decision-making is undoubtedly based upon the belief that any calculations that involve some degree of mathematical manipulation must be accurate. It remains the case, however, that life cycle costing is merely a guide to those making decisions, an additional piece of information that can be fed into the process of arriving at a final choice of design, component, maintenance programme and so on.

The actual mathematical techniques involved are straightforward and will be familiar to anyone who has undertaken investment appraisal; the primary technique involves discounting cash flows. All of the manipulations can be performed on modern micro-computers: indeed, we would recommend that this is the way in which the calculations should be performed in order to reduce the tedium and allow those doing the calculations to get on with the interesting questions that determine the final choice!

What life cycle costing does is look at the balance between initial and future expenditures. The basic idea is that spending a little bit extra now may well reduce expenditures in the future. In addition, more intangible benefits may flow from increased initial expenditures: in terms of improved aesthetic quality, reduced disruption during refurbishment or planned (or panic) maintenance, or increased income generating power of the building. Many of these intangible benefits are difficult to quantify in any objective way. Nevertheless they are important and should be allowed to influence the design process. We show how this can be done in a way that leaves the final choice to the design team and client: which is, of course, just where these decisions should reside.

We hope that readers of this book will find it both of practical use and of intellectual interest. More critically, we hope that we shall stimulate yet more of those working in the industry of the relevance and importance of life cycle costing. If these hopes are realised, then it is our belief that the industry will be better able to provide clients with the kinds of built environment that we all want: built environment that is efficient and pleasing. This can only be to the benefit of the industry.

Finally, a number of debts of gratitude must be acknowledged. First, we are grateful for the great assistance given by the British Ceramic Tile Council without whose help the project would not have been possible. Particular thanks are offered to the Executive Secretary for his support and guidance throughout the project. We also thank the British Ceramic Tile Council for permission to use data that could not have been generated without their considerable support. Secondly, the Editors have performed an invaluable task in helping us to produce what we hope is now a readable text. Also our thanks to Joan van Emden, John and Carol Jewell for their help in producing this book. Finally, but far from least, our secretaries in Leicester and Reading and our families have shown a patience that none of us have any right to expect.

Chapter 1 Introduction

There are numerous books and articles expounding the benefits of using a life cycle cost approach to evaluate complete buildings, elemental parts, systems, or the components and materials used in buildings. Life cycle costing is an obvious commonsense idea, in that all the costs arising from an investment decision are relevant to that decision.

Life cycle approaches cross disciplinary boundaries: the technique is equally applicable by building owners, surveyors, architects, engineers, contractors, specialist trade contractors, and materials manufacturers. For this reason the terms 'decision-maker' and 'analyst' have been used rather than references to any interested party or particular discipline. Tried and tested economic appraisal techniques are used, the main function of which is to facilitate the proper comparison of the various options available to the decision-maker.

We do not intend to re-state the debate about why life cycle costing techniques should be used. Our starting point is that life cycle costing has an essential role to play in decision-making, because the latter will be improved if it is based on the total costs of any investment decision rather than simply on the initial capital costs. A beginner's guide to life cycle costing has been provided in an earlier work by two of the present authors, Flanagan and Norman (1983). This book does offer some introductory concepts, but mainly describes a number of more advanced applications of life cycle costing in practice. As some of the approaches will be new to people in industry and the professions, we have concentrated mainly on evaluating one aspect of a building: internal finishes. This was selected for two reasons. Firstly, when applying life cycle techniques the decision-maker must understand the detailed technical issues relating to the components and materials: the life expectancy; the maintenance requirements; the performance characteristics; and so on, as well as the technical issues surrounding the application of life cycle costing techniques. In themselves these issues are very wide; hence the concentration on one important element of a building, the finishes. Secondly, we have been able to use information from a recent study, sponsored by the British Ceramic Tile Council, which gave access to data of high quality and detail.

In the first part of this book, Chapters 2 to 6, we have set out the theory of life cycle techniques and in the second part, Chapters 7 to 9, dealt with applications. Those readers experienced in life cycle costing may feel that the theoretical chapters can be skipped on first reading. But Chapter 2 is the only chapter that does not contain at least some new methodology, and even so, it discusses investment appraisal techniques in some detail.

Summary of a life cycle approach

There are four components: input; investment appraisal techniques; system application, and output, as shown in Fig. 1.1.

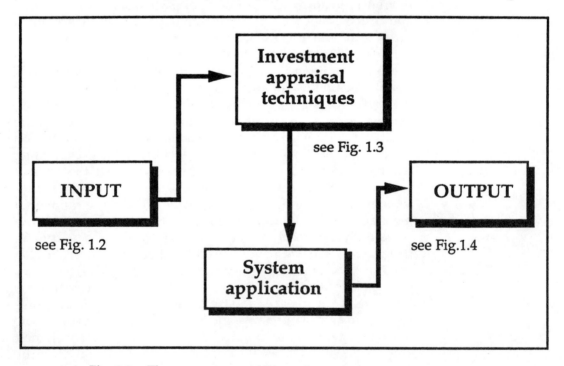

Fig. 1.1 The components of life cycle costing.

Feedback about buildings or part of a building's running costs and performance when it is in use

- Source of the data
- Reliability of the data (did we collect it ourselves, or was it from a published source?)
- Conversion of the data (into a meaningful structured format)
- How can the data be held on record?
- How old are the data (does it relate to one year, or over a number of years, how should it be averaged if required?)
- How old is the building and the main components? Does a block diagram of the building exist?
- What performance figures are available (how much electricity, gas, water, oil etc., has it used, how frequently is it cleaned?)
- How has the maintenance been managed and budgeted?
- What type of maintenance has been undertaken?
- What is the condition of the building (is there a condition survey available?)
- How should the data be updated to take account of the time differences (what index should be used?)
- How has the building been used/abused (what are the occupancy times, what are the types of activities?)
- If the building is rented are there any service charges payable?
- What is the superficial floor area?
- What is the glazed external window area?

Requirements for the proposed project

- Performance requirements (specified by client/designer team)
- Superficial floor and wall areas
- When is it anticipated the building will be occupied?

Fig. 1.2 Input.

Investment appraisal techniques

❑ Selection of an appropriate discount rate to take account of the time value of the money
❑ Selection of a period of analysis for the appraisal
❑ Deciding on how to handle future inflation (forecast, best guess, or hunch) for each item of expenditure
❑ Deciding on how the impact of taxation is to be handled
❑ Consider any risk premium required

System application

❑ Making assumptions about the periods of occupancy and how the building will be used (types of users)
❑ Deciding upon the methods of forecasting the running cost categories (energy and building performance models, using feedback data, obtaining data from manufacturers)
❑ Making forecasts about the life expectancy/obsolescence/decay of materials and components
❑ Forecasts about planned/preventative/unplanned/corrective maintenance, and about modernisation/refurbishment/adaption
❑ Forecasting salvage and residual values

❑ Evaluation of the costs and benefits
❑ Evaluating the risk and uncertainty
❑ Testing the sensitivity of the results (sensitivity analysis)

Fig. 1.3 Investment appraisal and system application.

Getting the results

❑ Should the results be displayed in tables or graph form?

❑ How will taxation affect the decision?

Fig. 1.4 Output.

The input is shown in Fig. 1.2; the investment appraisal techniques and the system application are shown in Fig. 1.3; and the output in Fig. 1.4. The information shown on these diagrams is discussed in detail in later chapters.

From theory to practice

The transition from understanding the theory of life cycle costing to practising it is not easy. The technique is heavily dependent both upon using running cost and performance data gathered from previous projects and upon forecasts of future events. Despite the progress in forecasting techniques, the fact remains that there is no infallible way to predict the future; forecasting is not an exact science. Nevertheless, it is an area of decision-making which cannot be left to go by default.

A constant criticism of life cycle costing is that it uses assumptions and forecasts which are no more than best guesses or hunches. There is nothing wrong with the 'best guess', as long as it enables the design team to make better decisions than would have been made otherwise. Note the use of the phrase 'better decisions'. Uncertainty surrounds the forecasting of an asset's physical or economic life, or future inflation or taxation rates. Who, in 1975, would have forecast annual inflation rates in excess of 20% within the following five years? And who, five years later in 1985 would have believed that annual inflation would be down to 5%? Forecasting will not guarantee correct decisions, but it will improve the basis upon which decisions are made. As has often been stated, 'It is better to be almost correct, rather than precisely wrong.'

We suggest that when active minds are applied to the best available data in a structured and systematic way, there will be a clearer vision of the future than would have been achieved by intuition alone. The effort will be justified even if it leads merely to the rejection of a few demonstrably wrong decisions.

What motivates clients?

Certain costs may be of much greater concern to some clients than to others, because clients differ in:

- their objective in building (e.g. short-term profit, long-term returns, satisfying public need);
- their timescale for the building (how long they expect to own it, how long before the purpose for which it is built will cease);
- their sources of capital (borrowed, retained profits, Government funding, grants);
- their source of revenues (e.g. income from letting, or use or production, Government), and
- their taxation position (if, or which taxes they pay, which costs can be claimed against tax, the rates at which tax is paid).

Time can be an important element in life cycle calculations. The client's objectives may be such that there are considerable gains to be achieved from having the building completed in a short time. This might preclude methods of construction with better maintenance characteristics. Whether the extra future costs are offset by the gain from early completion can be evaluated only by using a life cycle cost approach. Similarly, a slow construction programme may appear to offer a low cost. However, if interest payments, loss of business and so on are taken into account, the costs may be seen to be higher overall.

The importance of a total cost approach

Some building owners see running costs as uncertain, difficult to gauge and relatively unimportant compared to the running cost of their business as a whole. However, this blinkered approach, considering purely the initial capital cost, masks the considerable investment in their building stocks. Buildings are durable assets: they wear out, they become obsolete for a variety of reasons, but the buildings and the sites have residual and salvage values. All the time the building is in use it will be incurring running costs which cannot be ignored. These are not fixed but rather are variable costs and need to be budgeted in advance. Clients want to know the relative importance of the individual cost items in money terms, as a proportion of total costs.

There is a much more insidious and difficult obstacle to be overcome if a total cost approach is to gain wide operational acceptance. It remains the case in many organisations, in both the private and public sectors, that an impenetrable barrier is erected between capital and revenue budgets. Furthermore, it is often different people who are responsible for the capital and for the revenue budgets. Decisions to incur expenditure on the capital budget in the acquisition of durable assets – such as buildings and building components – take little or no account of the implications these decisions will have for those who operate the revenue budget – the eventual users of the durable assets. A total cost approach, almost by definition, treats capital and revenue budgets as an integrated whole. This integration allows the decision maker to move expenditure between what are, after all somewhat arbitrary budget headings. The justification for a particular choice of design or component will then be justified in terms of its *total* budgetary impact rather than its implications for one narrow aspect of the total budget of the organisation.

The consequences of having made a poor or wrong estimate about future performance at the design stage of a project takes a long time to come home to roost. There is no point in referring to past cost estimates when faced with maintenance expenditure today. No surveyor is likely to be held accountable for estimates made fifteen or twenty years previously that are now proven to be inaccurate. Few clients are likely to praise their consultants for spending more money today that 'might' show a saving in fifteen or twenty years time. The counter argument is when the forecasts of future maintenance expenditure are correct, few clients are prepared to face them. Crisis maintenance is a reality of everyday life. We can predict that certain types of mastics will start to fail from ten years old, yet frequently we do not make adequate provision for their renewal. Deferred maintenance has become a fact of life.

The picture is not all bleak. Many clients are increasingly interested in knowing more about the performance to be expected from their buildings, and in considering costs during the period of their interest in the buildings, as part of understanding the worth of their capital investment. They are breaking down the barriers between capital and revenue budgets. For example, in the USA there has been a mandatory requirement for life cycle costing to be carried out in the procurement process of all Federal building projects. The use of value engineering approaches in the USA helps in the implementation of life cycle costing at the design stage.

Clients are seeking appraisal techniques which can show the balance between future revenue expenditure and initial capital investment. The thrust is coming mainly from public-sector clients and owner occupiers, but speculative developers also are showing an increasing interest in achieving an efficient balance between current and future expenditures. In the appraisal of design and build submissions from contractors, where the design is based upon a performance specification, life cycle costing approaches have helped in the evaluation process.

A frequent criticism is that developers will not be interested in considering life cycle costs because running costs are the responsibility of the tenants. This has little, if any, foundation because tenants are interested in the service charges and running costs, and developers should want to provide the best value for money for their tenants. Furthermore, the capital market value of the building will be affected if tenants are unwilling to rent space because of the high running costs. In other words, the developer has to achieve a balance between current expenditure and future revenue.

Securitisation and unitisation of property

The current moves toward securitisation and unitisation of property will also call upon the use of life cycle costing techniques. Securitisation represents a new form of property ownership. It opens up the worlds of commercial property investment for the first time to the general public and potentially, allows property to be traded through the global securities market. In essence, individuals will purchase an equity share in a single property which is tradeable on the Stock Exchange.

The investors will be concerned about the value and condition of their building. Even though the building will be let on a full repairing lease there is no provision for the costs of refurbishment once the effects of obsolescence are felt. The investor is interested in the capital value of the asset and the income.

At the time of making the investment the investor will need to know the future expenditure on the building, both short term and long term. Plenty of warning would need to be given for any items of major expenditure. The property manager must maximise the performance of the building for the investor.

Life cycle costing can help in two ways. Firstly, it should be used at the investment stage to evaluate various investment opportunities. Secondly, during the life of the investment, it can be used to alert investors of future expenditure. With a property investment the property is the machine that generates the income and that investment has to be managed rather than merely monitored.

Putting total costs into perspective

The importance of considering total, rather than just initial, capital costs can be seen from the pie charts in Figs 1.5 to 1.8 inclusive. For each building type shown, both the initial capital cost of constructing the building and the subsequent running costs (such as annual and intermittent maintenance, cleaning, energy, security, and general and water rates) have been considered.

It can be seen that although in most cases initial capital cost is the largest single cost, in all cases it is under 50% of the total ownership costs of the building illustrated. The pattern of expenditure also varies between building types. In the elderly persons' home the running costs amount to 70% of the total cost, while for the primary school running costs are 55% of the total cost.

We have looked at the performance of the various buildings

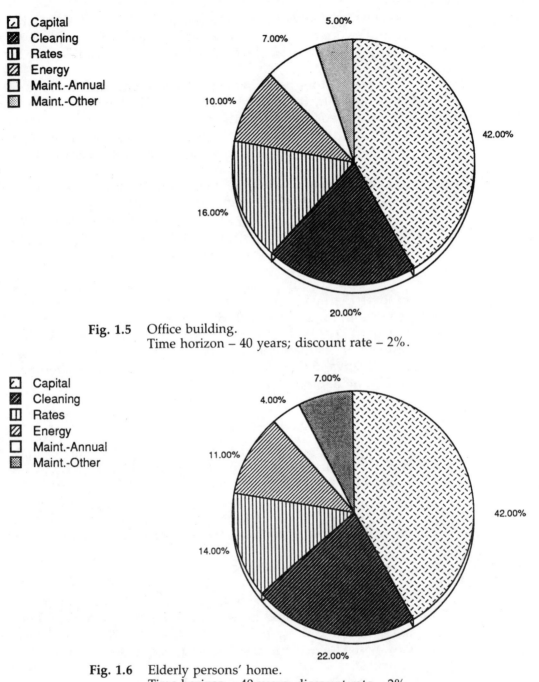

Fig. 1.5 Office building.
Time horizon – 40 years; discount rate – 2%.

Fig. 1.6 Elderly persons' home.
Time horizon – 40 years; discount rate – 2%.

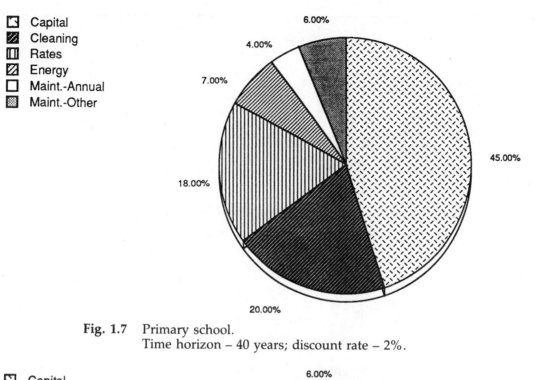

Fig. 1.7 Primary school.
Time horizon – 40 years; discount rate – 2%.

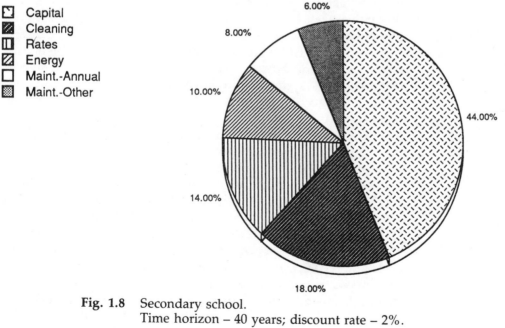

Fig. 1.8 Secondary school.
Time horizon – 40 years; discount rate – 2%.

Building type	Weekly usage (days)	Number of working days		Daily hours of occupancy
		in heating season	in whole year	
OFFICES	5	145	252	9
OFFICES	6	175	304	9
SHOPS	6	175	304	8.5
FACTORIES	5	145	252	9
RESIDENTIAL	7	210	365	24
HOTELS	7	210	365	24
HOSPITALS	7	250-365	365	24
EDUCATIONAL	5	160	200	7.5

Table 1.1   Building occupancy patterns.
(*Source*: Chartered Institute of Building Services Engineers – Energy Code Part 2.)

for the past ten years, projected expenditure over the period of analysis and converted the expenditure back to a net present value.

A key factor to be borne in mind is that buildings such as hospitals, elderly persons' homes, and hotels are generally in use 24 hours a day throughout the year, as is shown in Table 1.1. The hours of use and the occupancy profile of the building are important. Similarly, in a hospital building, essential maintenance cannot be deferred to ease pressures on the maintenance budget.

Although some of the costs occur on an annual basis, maintenance expenditure is both annual and intermittent. All building stock generates maintenance expenditure if its potential life expectancy is to be realised. Figures 1.5 to 1.8 show the distinction between annual and other maintenance. This emphasises the point that some maintenance expenditure can be delayed, for example internal decoration, some cannot be delayed without taking the building out of use, and some can be delayed with a future cost penalty. Lack of adequate maintenance causes an exponential deterioration as the maintenance needs increase: an unpainted window today means replacement of a rotten window tomorrow.

The figures give a snapshot of the percentages of expenditure in the various categories for the four building types over the same 40-year period of analysis. The total cost expenditure has not been shown, and much more information about the assumptions and forecasts would be needed in practice. For instance, no mention has been made of the residual value of the site and of buildings, or of the life expectancies of the fabric and components. Furthermore, the typical costs which have been deduced do not distinguish between good and bad performance or quality. The intention is simply to show that different building types have different expenditure profiles and a different relationship between initial capital investment and long-term running costs.

Looking at costs and benefits

The methodology traditionally used in life cycle costing incorporates basic discounting and financial appraisal techniques. The next logical step is to ally these objective techniques with rather more subjective techniques; for example, weighted evaluation matrices may be used for the initial selection of finishes based upon technical criteria.

Life cycle costing is commonly promoted as a cost-only concept, but all building expenditure, whether in the public or private sector, generates a value effect and this needs to be measured. The overall methodology and decision technique must, therefore, include a method to enable both the tangible *and* intangible costs and benefits associated with each choice to be assessed within an investment–appraisal framework. This is of particular importance because intangible costs and benefits have previously been found to play an important part in the selection process for finishes. However, no formal method has been available hitherto which could incorporate such costs and benefits explicitly.

A further reason for introducing intangible benefit appraisal derives from the fact that most clients have to operate within a capital cost constraint. 'I'd like it, but I can't afford it', is a fact of everyday life. It is a myth that the object of using life cycle costing techniques is to justify a higher initial capital cost. The intention is to evaluate the options: if clients cannot obtain or justify any additional capital expenditure, even though it might be economically viable in the long term, at least they have been given the options upon which to base their decisions, and shown what those decisions will actually cost in the medium and long term.

Some of the myths

Several myths have grown up around life cycle costing techniques:

Life cycle costing is concerned only with net present values and annual equivalent values which have no real meaning to everyday financial decisions.
Any life cycle costing system has to be tempered with reality. The client has to be confident that the figures being presented by the analyst represent reality. When considering net present values for capital expenditure, nobody is actually proposing to put away today the sums of money calculated for future expenditure. Any investment decision involves the time value of money. The use of discounting is simply a basis for looking at an exchange rate for 'tomorrow's money today'. In addition, the annual equivalent value of a particular option is the sinking fund the organisation should set up to pay for that option's installation, maintenance and subsequent replacement.

Life cycle costing cannot be valid in times of high inflation.
Untrue, because discount rates may be expressed in nominal (market) terms, where both the effects of general price inflation and the real earning power of money invested over time are reflected, or in real terms where only the real earning power of money is reflected and the effects of inflation are not included.

I cannot forecast three years into the future with any confidence, so why bother with a technique that looks at projects over 25 years?
We accept that in industry today, all the pressures on management are for short term gains. We have moved from the age of the stock market analysts looking at annual profits, to half time and interim performance. But buildings are long-term durable assets and must be considered as such; to ignore the future is folly. It is better to face the future and to try and make

forecasts rather than to consider it too difficult and to pretend events may never happen.

The technique is not appropriate to refurbishment projects.
Life cycle approaches are equally applicable to new build, refurbishment, or decision-making about an individual component.

Life cycle costing must include the whole life of a project until disposal.
There is confusion about the life cycle of the component and the period of analysis: they are separate issues. If a shop is intended to be in use for only three years then that is a perfectly reasonable period of analysis; the shopfitting components may well have a design life of ten years or even more.

Initial good intentions to apply life cycle principles to design almost always fail under the real life pressures of meeting deadlines and budgets.
There will always be pressures to meet deadlines and budgets, no matter how good the design team or how straightforward the project. Life cycle costing has to become part of the overall system and not merely a tool for effective decision-making. Life cycle costing is a way of thinking and making people take responsibility both at the design stage and at the management stage of the asset in use.

Each building is a unique product and therefore feedback about the running costs from any one building is not likely to be relevant to other buildings.
This is based on the fallacy that the only way to predict future running costs is to concentrate on past performance. Manufacturers and component designers have a role to play in forecasting the performance. Similarly, simulation models, such as those used for forecasting the energy requirements of buildings, are now becoming more reliable.

When are life cycle techniques useful?

- As an evaluation technique helping to choose between competing options, whether these relate to a complete building, a system, or a material.
- As a basis for predicting future running costs.
- As a management tool to ensure that the facility is being used

effectively and that maximum value for money is being obtained. For example, has a particular element been unduly expensive? Should a different type of material, system or structure have been used? What types should be avoided and which should be used by the organisation when a similar type of building is to be constructed under similar circumstances?

- As a basis for budgeting for future expenditure. All clients must manage their assets within an annual rolling budget, even the householder has to ensure that the annual fuel bill for heating and hot water is within his budget. Clients will therefore have a short and a long term strategy. Life cycle costing helps to identify and assist in balancing future expenditure on the asset.
- As a means of considering total cost rather than merely initial capital cost.

The terminology

If a total cost approach using life cycle costing is more than just a good idea, several issues must be confronted in making that idea a reality. The first is the terminology, the most essential of which is set out in the Glossary on page 175.

Next, we need to consider the various discounting techniques essential to the calculation of life cycle costs: this is done in Chapter 2. It will become obvious that life cycle costing is potentially expensive in its data requirements and Chapter 3 discusses how these requirements may be met. Uncertainty is endemic to all attempts to deal with the future, but also can be used to improve decision-making and Chapter 4 shows how this can be achieved. Finally, intangible considerations are important: these are discussed in Chapter 5, while the remaining chapters concentrate on applying these various techniques to a specific example.

Chapter 2 *The application of life cycle costing: techniques and issues*

The theories developed in this and the following chapters are illustrated later by applying them to the evaluation of building finishes. For consistency, therefore, we have referred throughout to finishes, but it is a simple matter to translate this discussion to any other building component.

This chapter develops a number of the quantitative techniques on which life cycle costing is based. These techniques are intended to guide the choice between options; for example, the choice of floor or ceiling finish in a particular functional space. We are seeking a method for choosing, on some objective and quantifiable basis, between the various finishes that satisfy the technical and other criteria. This involves the application of some kind of investment appraisal to the choice of finishes.

Any systematic review of the literature on investment appraisal methods will reveal a bewildering variety of suggested criteria on which to base decisions, but the most basic requirement to be satisfied is that the method must be capable of ranking the various options according to some defined measure. In this respect, a decision-making method based on minimum capital cost is certainly a candidate as a means of investment appraisal. By the same token, however, so is one based on minimising running and other recurrent costs. Since the installation of finishes will involve the owner in both capital and recurrent costs, what is needed is a technique of investment appraisal that takes both sets of costs into account.

The techniques

Pay-back

Simple method
This is one of the most tried and trusted of all methods because of its simplicity in calculation and ease of interpretation. Pay-back is simply a measure of the time required to return the initial investment. The ranking criterion is that the preferred option has the shortest pay-back.

In applying pay-back to the choice of finishes, one problem has to be overcome. The 'traditional' application of pay-back looks at projects with an initial capital outflow followed by subsequent cash inflows. By contrast, in choosing between competing finishes, all cash flows will be outflows (with the possible, but minor, exception of any residual value at the end of the life of the project).

In order to overcome this problem, an incremental approach must be used, such as that illustrated in Table 2.1. Assume that there are two competing finishes, A and B, with cash flow profiles as in Table 2.1. The hypothetical project B–A is

Initial Cost	Finish A (£30,000)	Finish B (£70,000)	B - A (£40,000)
	£	£	£
Cash Outflow			
Year:1	(17,500)	(2,500)	15,000
2	(17,500)	(2.500)	15,000
3	(17,500)	(2,500)	15,000
4	(17,500)	(2,500)	15,000
5	(17,500)	(2,500)	15,000
Pay-back for B-A			2.7 years

Table 2.1   Application of pay-back method.

constructed by choosing as base the finish with the lowest initial capital costs (finish A). The pay-back for this hypothetical project is calculated and compared with the minimum standard required. If the pay-back period on the incremental investment for the finish with the more expensive initial capital costs is less than the minimum standard, the more expensive alternative should be adopted. In the example in Table 2.1, if the minimum standard is three years or more, finish B should be adopted in preference to finish A.

The advantages of the pay-back method are, firstly, that it is quick and simple to calculate once the cash flows for the various candidate finishes have been forecast. Secondly, the results are easily understood by management. Thirdly, the underlying rationale for the method is intuitively appealing, emphasising as it does the speed of return. This can be an important factor when cash flow is a limiting constraint on the client, and goes some way to mitigating the effects of risk inherent in forecasts of future cash flows.

Despite these apparent advantages, there are good reasons for rejecting the pay-back method as a suitable means of evaluation. The first is the ambiguity of what is meant by the 'initial capital cost'. Consider the options in Table 2.2: these

	Incremental Cash Flows	
Year	Finish C-A £	Finish D-A £
0	(10,000)	(5,000)
1	5,000	1,000
2	5,000	(5,000)
3	5,000	3,000
4	(2,000)	3,000
5	4,000	4,000

Table 2.2   Definition of initial cost.

patterns may arise if, for example, the cycle of intermittent replacement costs differs between finishes A, C and D.

What are the initial capital costs in the two incremental finishes C–A and D–A? Is incremental finish C–A's initial capital cost £10 000 or £12 000? Is incremental finish D–A's initial capital cost £5 000 or £10 000? Clearly, a managerial view can be taken of the appropriate measure of these costs, but the decision rule will be ambiguous. Such ambiguity presents serious problems. It can never be clear that the rule applied has not been chosen in order to justify the desired decision rather than the correct one. Any decision rule with such inherent ambiguity must be treated with extreme caution and mistrust.

The second problem relates to the setting of maximum acceptable pay-back. What defines this decision-making criterion? In short, there is, again, the potential for ambiguity in that the criterion upon which choice of finish is based is not objective; rather, it is based on nebulous and ill-defined concepts such as company or industry 'norms'.

Thirdly, and perhaps most fundamental of all, the pay-back method ignores all cash flows outside the pay-back period. The incremental finishes E–A and F–A in Table 2.3 illustrate this problem. Finish F would be preferred to finish E, and would be chosen if the maximum acceptable pay-back exceeded two years, but the incremental cash flows in years 3, 4 and 5 cast some doubt on the correctness of this decision advice.

It has often been suggested that a further reason why the pay-back method should be rejected is that it fails to take account of the time value of money, but this limitation is easily overcome by using discounted pay-back as a criterion.

Discounted method
To see how discounted pay-back might work, consider the time value of money. This means that a given sum of money has a different value depending upon when it occurs in time. For example, suppose you were asked to choose between receiving £100 today or £100 in one year's time. Intuitively, most individuals would choose to take the money today. This is not solely because of inflation or risk, although of course these are important considerations. Rather, it is because if you are given the money today you could choose to invest it and so earn interest. Your £100 would then grow over the year to £106 if the

	Incremental Cash Flows	
Year	Finish E-A £	Finish F-A £
0	(20,000)	(20,000)
1	2,000	10,000
2	4,000	10,000
3	6,000	0
4	16,000	1,000
5	16,000	0

Table 2.3 Cash flows outside the pay-back period.

	Incremental Cash Flows	
Year	Finish G-A £	Finish H-A £
0	(5,000)	(5,000)
1	4,000	2,000
2	3,000	3,000
3	2,000	4,000
4	1,000	1,000
5	1,000	1,000

Table 2.4 Discounted pay-back.

interest rate were 6%. What this implies is that if you were, instead, offered the choice between £100 today and £106 in one year's time, you would be indifferent between the two sums. Thus, £100 today and £106 in one year are in some sense equivalent: the present value of £106 in one year is £100 (or the future value of £100 in one year is £106), assuming of course a 6% interest rate.

Applying the concept of the time value of money to future cash flows indicates that the present value of the cash flows is less than their forecast (i.e. future) values. The simple pay-back method would not distinguish between the incremental finishes G–A and H–A in Table 2.4: for example, either would qualify with a maximum acceptable pay-back of four years. Discounted pay-back would result in a preference for G–A, since the large net cash inflows occur earlier on G–A than H–A.

Despite the undoubted drawbacks of the pay-back method, simple or discounted, it has exhibited a remarkable resilience and popularity. Undoubtedly this is because of its ease of calculation and interpretation. The real problem lies not in the basic concept of the pay-back method, but in the way it is used. Pay-back gives useful information on the competing options, but should not be used to give decision advice. It cannot be considered a reasonable investment appraisal method, precisely because it ignores all cash flows outside the pay-back period. At most, it should be used as an initial screening device before the application of more powerful methods of appraisal.

The time value of money
Any acceptable investment appraisal technique must exhibit two properties:

* it should take account of all cash flows associated with the investment throughout the period of analysis;
* it must make proper allowance for the time value of money.

Simple pay-back fails both tests, and discounted pay-back fails the first.

There is a third criterion that should also be required of an investment: the return on the investment should be not less than the market rate of interest (allowing for uncertainty, which is referred to later). Pay-back does not meet this criterion, since it uses 'time' rather than 'rate of return' to assess project desirability. In developing an investment appraisal technique (discounted cash flow or net present value) that meets all three criteria, it is necessary first to explain in more detail what is meant by the time value of money.

We have seen already that the value of a sum of money depends upon the time it is received or expended. If I need to spend £105 (at today's prices) on maintenance of the ceiling finishes next year, what sum should I earmark today for this purpose? Assuming that money can be invested at a real rate of interest of 5%, £100 needs to be committed today. In other words, with an interest rate of 5%, an expenditure of £105 next year is equivalent to an expenditure of £100 today.

This shows that with an interest rate of $r\%$, an expenditure of £P today is equivalent to an expenditure of £P $(1 + r)$ in one year's time. The expenditure next year could be referred to as a terminal expenditure. Thus the terminal expenditure in one year's time, T_1, equivalent to a present expenditure of P is:

$$T_1 = P\,(1 + r) \tag{1}$$

Put another way, the present value of the expenditure T_1 in one year's time is:

$$P = T_1/(1 + r) = T_1\,(1 + r)^{-1} \tag{2}$$

When it is forecast that a particular floor finish will require regular maintenance, at today's prices, of £100 per annum and again the interest rate is 5%, what sum must be committed today to meet these expenditures? The present value of the expenditure in one year's time is given by the equation:

$$P = T_1/(1 + r). \tag{3}$$

Thus the £100 spent in year one has present value: £95.24 = £100/1.05.

Now consider the expenditure of £100 in year two. By the same argument, that expenditure would be covered by committing £95.24 in year one and £95.24 in year one is covered by committing £90.70 = £95.24/1.05 today. That is, the present value of the expenditure in two years' time is:

$$£90.70 = \frac{£100}{(1.05)(1.05)} = £100/1.05^2 = £100 \times (1.05)^{-2}$$

This argument can be extended *ad infinitum*. The present value of a sum of £100 expended in n years' time with market interest rate 5% is:

$$P_n = 100/(1.05)^n \tag{4}$$

The total present value of the regular maintenance expenditures over N years is then:

$$PV = P_0 + P_1 + P_2 + \dots + P_n + \dots + P_N \tag{5}$$

$$= 100 + \frac{100}{1.05} + \frac{100}{(1.05)^2} + \dots + \frac{100}{(1.05)^n} + \dots + \frac{100}{(1.05)^N}$$

More generally, the present value of an expenditure T_n in n years' time with interest rate r is:

$$P_n = T_n/(1 + r)^n \tag{6}$$

and the present value of the annual maintenance cash flows is:

$$PV = T_0 + \frac{T_1}{1+r} + \frac{T_2}{(1+r)^2} + \dots + \frac{T_n}{(1+r)^n} + \dots + \frac{T_N}{(1+r)^N} \tag{7}$$

The procedure being applied in calculating this present value is called discounting, and the interest rate is referred to as the discount rate.

It must be emphasised that the present value of a future sum is not just a strange piece of arithmetic, but has a specific economic meaning: it represents the sum of money which, if set aside today, will exactly cover the future expenditure. In other words, it represents a price.

For example, suppose:

- you know you will need to replace 10% of your floor tiles in two years' time, at a cost in today's prices of £110.25;
- the Government has conquered inflation and that the interest rate is 5%;
- you want to enter into a contract today with the supplier such that, in exchange for an agreed sum of money today, he will supply the replacement tiles in two years' time.

In this certain, inflation-free situation, you would be foolish to offer more than £100 and he would be foolish to accept less. If, instead of offering him, say, £105 you had placed this sum on deposit, in two years it would have been worth £115.76 (£105 × $(1.05)^2$ = £115.76) whereas you would have received goods then worth only £110.25. On the other hand, the supplier would be irrational to accept less than £100. If he were to accept £95, then this would grow to £104.73 (£95 × $(1.05)^2$) in two years' time, whereas he would be committed to supplying goods then worth £110.25.

This argument indicates that the present value of a future sum has a clearly-defined economic meaning: it is the present exchange value of that future sum, given the interest rate at which it is calculated. A further implication of this analysis is that the present value of a future sum is less the further away in time the sum is due.

We conclude that the value of an expenditure is determined by when it is made: money has time value. In addition, expenditures at different points in time can be compared. Is it better, for example, to spend £100 on maintenance in year two or £106 in year three? If the £100 is invested in year two, it will grow to something greater in year three, but if the interest rate is less than 6%, the £100 invested will grow to less than £106 in year three; the money would have been better spent on maintenance. Conversely, if the interest rate is more than 6%, it would be better to invest the money and postpone expenditure on maintenance to year three.

Equally, the £106 can be discounted to year two, to find its value in year two. This will result in exactly the same answer: if the discount rate is less than 6%, the discounted value of £106 will exceed £100, and expenditure on maintenance is cheaper in year two, while with a discount rate of more than 6%, maintenance is cheaper if postponed to year three.

This shows that in order to find the total value of a series of expenditures occurring at different points in time it is incorrect simply to add them up: this implicitly assumes that the timing of the expenditures is irrelevant. It is, therefore, meaningless to talk of the value of such a time-stream of expenditures without also specifying the date of the valuation. If any meaning is to be attached to the concept of the total value of expenditures distributed over time, they must first be expressed in a common unit, as if they all occurred at the same point in time. The price implicit in the time value of money, and the process of discounting, allow this conversion to be effected.

An optimal investment criterion: net present value

The investment appraisal technique referred to as the net present value (NPV) technique satisfies the three necessary criteria already described:

- it uses all available data
- it takes account of the time value of money
- it generates a return at least equal to the market rate of interest.

Why is this so?

Firstly, we shall define what is meant by the net present value of a particular choice of finishes. Given

- estimated initial and future costs, where C_t^i is the estimated cost for finish i in year t
- a discount rate of r
- a period of analysis of T years

the net present value of finish i is:

$$\text{NPV}_i = C_0^i + \frac{C_1^i}{1+r} + \frac{C_2^i}{(1+r)^2} + \ldots + \frac{C_t^i}{(1+r)^t} + \ldots + \frac{C_T^i}{(1+r)^T} \quad (8)$$

$$= \Sigma_{t=0}^{T} \frac{C_t^i}{(1+r)^t}$$

The optimal decision rule also follows naturally. Mutually-exclusive investments are being compared: the choice of one finish in a particular environment excludes the choice of any other finish. The optimal choice of finish is, therefore, the finish with the lowest net present value. In other words, choice is determined by ranking the options in ascending order of their net present value.

This all seems very straightforward, but why does this decision rule hold true? What does it mean? Return to the interpretation of net present value: from looking at equation (8), it is clear that the net present value of finish i is the sum of the present values of the annual cash flows associated with finish i. The discussion of the time value of money indicated that the present value of a cash flow forecast for year t is the sum that must be committed today to cover that cash flow. This gives a very simple and appealing interpretation of net present value: it is today's price of the total investment to which the decision-maker is committed as a consequence of a particular choice of finish. Alternatively, it is the value of the sinking fund that would have to be established today in order to cover all of the cash outflows associated with a particular finish (initial installation, maintenance, intermittent replacement, and so on) over the relevant period of analysis.

The components of net present value
Calculation of net present value according to equation (8) has three essential elements as input:

- estimates of costs
- the discount rate
- the period of analysis

Estimated costs will be dealt with in some detail in subsequent chapters when sensitivity analysis is considered.

The discount rate
In the examples above in which a discount rate is used, reference has been made to costs at today's prices, and an inflation-free discount rate. This implies that there might be some connection between the calculation of the discount rate

and the relevant inflation rate: a distinction must be drawn between discount rates and interest rates.

An interest rate is made up of two components, the time value of money and the effects of inflation. Since inflation has become a significant factor to be considered when predicting future costs it should be taken into account within the discount rate. The following equation does this:

$$d^1 = \left[\frac{(1+d)}{(1+i)} \right] - 1 \tag{9}$$

where

d^1 = net of inflation discount rate (real discount rate);

d = interest rate including inflation (nominal discount rate);

i = inflation rate.

It must be stressed that, in every option considered, the costs must be calculated on the same basis. For example, it should be explicitly stated whether cash flows take into account the effects of inflation, and, if so, this should extend to all the options under consideration so that they may be compared on a 'like for like' basis. Equally, in performing the discounting calculations it is essential that cash flows and discount rates are chosen on the same basis. If the cash flows exclude inflation (expressed in today's prices) the real (inflation free) discount rate of equation (9) must be used.

Choice of the discount rate
Choosing the right discount rate will depend on the circumstances and objectives of the client. However, in most cases the choice depends upon whether the client is financing the project through borrowed money, or from capital assets.

In the first case the appropriate discount rate must be equivalent to the actual cost of borrowing the money. If, however, the project is to be financed from capital assets (for example, retained income, funds from the issue of shares, or debt capital) the discount rate must be determined by the current and future rate of return for that client's particular

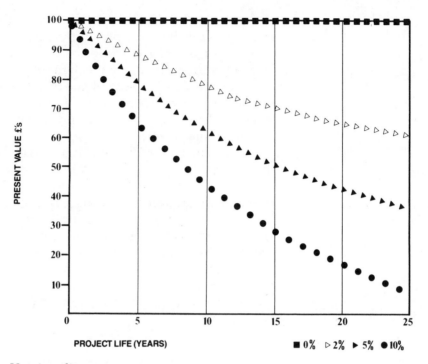

PRESENT VALUE £'s

PROJECT LIFE (YEARS)

■ 0% ▷ 2% ▶ 5% ● 10%

Fig. 2.1   Varying discount rates.
Present value – £100.

industry and, ultimately, by the best alternative use of such funds.

The discount rate clearly is one of the critical variables in the analysis, in that the decision to proceed with a project will be crucially affected by which discount rate is chosen. Too high a discount rate will bias decisions in favour of short-term low capital cost options, while too low a discount rate will give an undue bias to future cost savings. Figure 2.1 shows the effects clearly.

Other things being equal, the higher the discount rate used, the less impact future running costs will have upon the investment decision.

Benefit-cost ratios

One of the problems with net present value as an investment appraisal technique is that decision-makers find it difficult to interpret. In the previous section two intuitive meanings were given to NPV, but interpretation is probably one of the major reasons for the resistance to the use of NPV, not only in the building industry, but in a much more general context. In order to overcome this problem, some analysts have suggested that the internal rate of return (IRR) should be used to rank optional projects. This is defined as the discount rate that gives a net present value of zero.

One problem of applying IRR to choice of building finishes is that all cash flows on a particular choice of finish are negative (cash outflows) making calculation of IRR impossible: the same problem encountered with the pay-back method. This can be overcome by using the same incremental approach discussed above in the context of pay-back. Given two finishes, A and B, the cash flows on the incremental project B–A will consist of a mix of negative net cash flows – which can be denoted C_t – and positive net cash flows – denoted R_t. This is illustrated in Table 2.5 where, by convention, if $C_t = 0$ then $R_t \neq 0$ and vice versa.

The IRR is calculated from the equation:

$$\text{IRR} = i : \Sigma_{t=0}^{T} \frac{R_t}{(1+i)^t} - \Sigma_{t=0}^{T} \frac{C_t}{(1+i)^t} = 0 \qquad (10)$$

	Finish A	Finish B	B - A R_t	C_t
	£	£	£	
Initial Cost	(1,000)	(1,200)	200	
Cash Outflow				
Year 1	(600)	(500)	100	0
Year 2	(600)	(500)	100	0
Year 3	(600)	(500)	100	0

Table 2.5   Internal rate of return.

The reader is left to check that in this example IRR = 23.7%. Choosing finish B over finish A would be sensible provided the required rate of return is less than 23.7%.

IRR has an obvious appeal because it is presented as a percentage, with an obvious interpretation. There are, however, many reasons why IRR should not be used as a method of ranking competing investments. These need not be considered in any detail here; they are discussed in Bromwich (1977), and Quirin and Wiginton (1981). One fundamental problem with IRR, for instance, is the implicit assumption that all net cash inflows on a particular investment choice can be reinvested at the IRR: in the example in Table 2.5, IRR assumes that the net cash savings offered by finish B in years 1, 2 and 3 can be reinvested at 23.7%.

It is, however, possible to develop an investment appraisal method that bears at least a family resemblance to IRR: this is the benefit–cost ratio (BCR). The BCR for a particular project is the ratio of the present value of future benefits at a specified discount rate, to the present value of the future costs discounted at the same rate. By definition, therefore, it can be applied only to incremental projects. Using the notation of Table 2.5 and equation (10), the BCR for the incremental finish B–A is:

$$\text{BCR} = \frac{\Sigma_{t=0}^{T} R_t/(1 + d)^t}{\Sigma_{t=0}^{T} C_t/(1 + d)^t} \tag{11}$$

As can be seen, BCR takes account of all cash flows, and the time value of money. It might be expected, therefore, that there will be some relationship between BCR and NPV. This is shown in the example in Table 2.6.

The difference in NPV between finish M and finish L is, in fact, closely related to the BCR of the incremental finish M–L through the formula:

$$\text{NPV}_M - \text{NPV}_L = \frac{\text{BCR}_{M-L} - 1}{\Sigma_{t=0}^{T} C_t/(1 + d)^t} \tag{12}$$

Recall that $\text{NPV}_M - \text{NPV}_L$ will be positive if the present value of the total cost of finish M is less than the present value of the

Table 2.6 Benefit cost ratio and net present value.

	Finish L	Finish M	Discount Factor @ 5%	Present Value L	M	M - L	Present Value Rt	Ct
Initial Cost	(6000)	(7000)	1	(6000)	(7000)	(1000)	-	1000
Cash Outflow								
Year:								
1	(1000)	(600)	0.952	(952.0)	(571.2)	400	380.8	-
2	(1100)	(600)	0.907	(997.7)	(544.2)	500	453.5	-
3	(1300)	(700)	0.864	(1123.2)	(604.8)	600	518.4	-
Net Present Value				(9072.9)	(8720.2)			
Net Benefits discounted							1352.7	
Net Costs discounted								1000
Net Present Value M - L					352.7			
Benefit Cost Ratio M - L							$\frac{1352.7}{1000} = 1.353$	

total cost of finish L. It then follows from equation (12) that if finish M is the lower cost option, $NPV_M - NPV_L$ will be positive (offer a cost saving) and BCR_{M-L} will be greater than unity. Two equivalent decision rules are obtained in choosing between the two alternative finishes:

- choose the finish with the lower NPV
- choose the incremental finish for which BCR exceeds unity.

In addition, the BCR can be expressed as a rate of return in excess of the discount rate. In the example of Table 2.6 the additional initial capital cost of finish M offers an enhanced return of some 10.6% ($1.106^3 = 1.353$) above the discount rate.

In other words, the BCR can be used to express the cost savings of the lowest total cost option as a rate of return, thus gaining some of the benefits of IRR in its appeal to the intuition of decision-makers.

One cautionary note must be sounded. Where there are three or more finishes to be compared, NPV gives a very simple rule: choose the lowest NPV. BCR does not: a series of paired comparisons have to be made. It could be, for example, that with three finishes, A, B and C:

(1) $BCR_{B-A} > BCR_{C-A}$ *and*
(2) $BCR_{C-B} > 1$.

Finish C should be chosen.

Annual equivalent value: investment with unequal lives

As long as attention is confined to choices between building components such as finishes, it is reasonable to assume that the same period of analysis will be used in the comparison. Nevertheless, there remains something of a problem if the natural replacement cycle of a particular component is not an exact multiple of the period of analysis. For example, what should be done if a component or finish should be completely replaced every 10 years, and the period of analysis is 25 years? The third finish would still have five years of useful life at the end of the analysis.

This can be handled by attributing a residual value to cover the remaining years of useful life (entered as a revenue, i.e. a positive contribution) in the analysis.

This can be somewhat arbitrary. An alternative method which has been advocated to solve this problem is the annual equivalent value. Suppose that there are two finishes, A and B, with natural replacement cycles n_A and n_B years respectively. Calculate the NPV of the two finishes over their respective expected lives:

$$NPV_A = \Sigma_{t=0}^{n_A} C_t^A/(1 + d)^t \qquad (13)$$

$$NPV_B = \Sigma_{t=0}^{n_B} C_t^B/(1 + d)^t \qquad (14)$$

These NPVs cannot be compared directly, since they refer to investments with differing lives. To obtain comparable measures, first identify from annuity tables the present value of an annuity of n_A years at $d\%$, and of an annuity of n_B years at $d\%$, denoted AN_A and AN_B respectively. If, for example, $n_A = 5$ years, $n_B = 8$ years and $d = 5\%$ then

$$AN_A = 4.3295 \qquad AN_B = 6.4632$$

The NPVs can now be expressed as annual equivalents as follows:

$$\begin{array}{l} \text{Annual equivalent of} \\ \text{investment A} \end{array} = S_A = \frac{NPV_A}{AN_A} \qquad (15)$$

$$\begin{array}{l} \text{Annual equivalent of} \\ \text{investment B} \end{array} = S_B = \frac{NPV_B}{AN_B} \qquad (16)$$

If, for example, $NPV_A = £5200$ and $NPV_B = £6100$, with $n_A = 5$ years, $n_B = 8$ years and $d = 5\%$, then

$$S_A = \frac{5200}{4.3295} = £1201.06$$

$$S_B = \frac{6100}{6.4632} = £943.80$$

and finish B should be chosen.

The annual equivalent value is the regular annual cost (annuity) that, when discounted, just equals the NPV of the investment. Choosing the investment with the lowest annual equivalent value must, therefore, minimise total cost.

Again, a cautionary note is necessary. The annual equivalent value method can only be used when there is a natural replacement cycle over which the finish is completely replaced; it does not refer to a regular maintenance cycle. There is also the assumption that the same finish will be used in each replacement cycle – ceramic tiles will be replaced by ceramic tiles, carpet by carpet, and so on. Effectively, the annual equivalent value method is merely a short cut method for comparing investment programmes consisting of consecutive identical investments.

The issues

Timescale

The timescale over which the life cycle costs are calculated is fundamental. Two different timescales are involved:

- the expected life of the building, the system, or the component
- the period of analysis.

It is important when carrying out any form of life cycle costing to differentiate between these two timescales, since there is no reason to believe that they will be equal: for example the recommended period of analysis for Federal buildings in the USA is 25 years, considerably less than any reasonable building life.

Expected life
Two considerations are of primary importance in determining expected life:

- the expected service life of the building (e.g. structure and fabric);
- the expected service life of each component (e.g. floor finishes).

The expected or useful life of any product is determined by a variety of factors, some environmental and some non-environmental. Any material or component will deteriorate because of environmental factors such as radiation (solar and thermal), temperature ranges, water (rain, condensation, snow, ice), air contaminants, biological factors (micro-organisms, fungi, bacteria) and stress factors (physical action of wind, hail). The non-environmental factors are generally the stresses that are imposed by humans in their various activities of living, working and playing. Examples are permanent loading, fatigue loading, impacts, abrasion, chemical attack, normal wear and tear, and abuse by the user. More specifically, stresses on floor coverings in a building are caused by:

- abrasion from foot traffic
- indentation from static furniture
- indentation from moving furniture
- impacts from dropped goods
- water for cleaning and spillage
- chemicals for cleaning.

The levels of stress and frequency of stress will vary with the type of use, and the number of occupants.

Various expressions are used to define the life of a product. The most common are:

Acceptable life   This depends upon who is defining what is acceptable. The organisations that lend money on buildings require that the building should survive the period of the loan.

Average life   The time to 50% failure. Usually this can be established only after the event. The points to remember are that two materials with the same average life may have a totally different failure time distribution about the average life, and that the same material used in different locations may give different average lives.

Minimum life   In some situations it is not possible to give an accurate assessment of a product's life because of a lack of data. Many material testing agencies, such as the British Board of Agrément (BBA), will stipulate that the product will last at least x years, which is the minimum life. The true life of the product may be much greater.

Design life   Some specifications will call for a design life of a product. A good example is the Department of Transport, who have a requirement for a design life of 120 years for bridges and major works on motorways, subject to regular specified maintenance being undertaken.

Expected service life   The term used for life cycle costing exercises which use expert judgement to evaluate an expected life which is a balance between the minimum life and the design life. Due account must be taken also of the obsolescence factors mentioned below, whether technological and/or social.

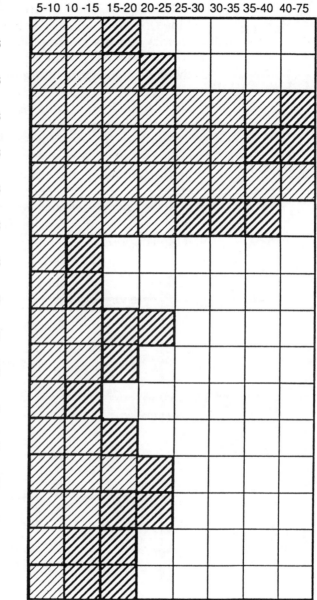

Fig. 2.2 Average life expectancies in years. Hatching indicates range dependent upon system or type chosen.
(*Source*: UK Property Report 1985, Richard Ellis.)

The expected service life of the building can be defined as the timescale over which the building structure and fabric can be maintained in an acceptable physical condition. Some of the component parts will need replacement from time to time during the life cycle of the building, just as they do with a car, for instance. A good example from everyday life of historical data on the lives of components in use in commercial buildings is shown in Fig. 2.2.

This discussion is further complicated by the fact that the expected life of (for example) a building finish will be affected by the frequency of maintenance. This is illustrated in Fig. 2.3, where (a) and (b) illustrate typical deterioration rates for building structures and components. Figure 2.3(b) illustrates how expected life can be extended by repair/refurbishment/modernisation.

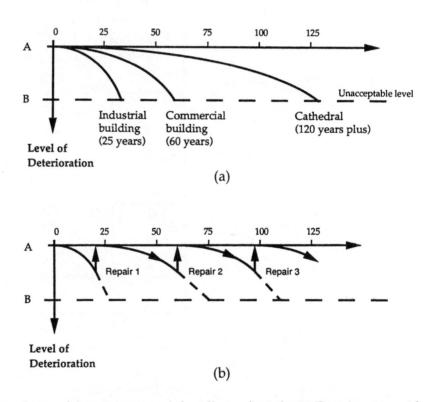

Fig. 2.3   Rates of deterioration and the effects of repair. (a) Deterioration with age. (b) Repair/refurbishment/modernisation.

Period of analysis

The timescale over which the life cycle cost will be considered will not necessarily be the same as the expected physical life of the building or the individual building component. The period of analysis can be defined by the expected period of use that either the building owner or user will require from the building, that is, how long he will be involved in the decision either to build, or to occupy it.

A number of factors relevant to the use of the building by an owner or occupier will assist in determining the period of analysis. These may be summarised as follows:

- physical deterioration
- economic obsolescence
- functional obsolescence
- technological obsolescence
- social obsolescence
- locational obsolescence
- legal obsolescence
- aesthetic and visual obsolescence (fashion/image obsolescence)
- environmental obsolescence.

Economic obsolescence is the most common form, and arises from the change in the best use of the land on which the building is sited; it is often a result of the enhancement of the land value, rather than physical deterioration of the building structure.

The impact of these factors will vary from client to client, and from component to component. For example, a major UK retailing chain uses a period of analysis of five years for shop fittings, but rather longer for air-conditioning systems.

The appropriate choice of the period of analysis can be made only in the individual context. The outstanding issue is how to resolve differences between the period of analysis and the expected building or component life. A simple example will serve to illustrate how this is done. A client commissions a 'high-tech' office building that has an expected physical life of 60 years. It is also estimated that the building will need retrofitting in 25 years' time. How should this be interpreted? If the building is expected to last for a period of 60 years but the owner foresees that perhaps in terms of his 'company image' or future running costs the building will become economically obsolete for his use in 25 years' time, the period of analysis will become 25 years. The residual value, to be included within the analysis, will be the building's expected market value or opportunity cost from 25 years to 60 years. After 25 years, if a decision is made to retrofit the building, a completely new analysis will begin.

One point worth noting in this context is clear from Fig. 2.1, which shows that when discounted the present value of a particular cash flow is less, the later in the time horizon the cash flow occurs. Thus, cash flows occurring beyond about 25 years are unlikely to make any significant difference to the ranking of options.

Obsolescence and deterioration
The types of obsolescence have been discussed previously, but often the implications are not fully appreciated. Obsolescence may be defined as the value decline that is not caused directly by use or passage of time, but a distinction should be drawn between obsolescence and deterioration. This is particularly important because when preparing company accounts certain types of plant and equipment can be depreciated as a result of deterioration. Depreciation is taken into account because physical assets wear out over time, which has an impact both upon the market value of the assets and upon the need to make provision for the cost of replacement.

Physical deterioration of a building and its components is a function of its use and time. The deterioration can be controlled to some extent by the provision of good-quality components and high levels of maintenance. The rate of deterioration can be modelled in some way as is shown in Fig. 2.4.

The control of obsolescence is much harder to achieve, since it is influenced by uncertain future events and is irregular in nature. Deterioration results in an absolute loss of utility, while obsolescence produces a relative loss of utility. For instance, a building system may be in as good as new condition but a superior system may impart total obsolescence: a good example of this has happened with the advent of sophisticated word-processing equipment, which has resulted in the scrapping of many typewriters.

Obsolescence is particularly important for commercial

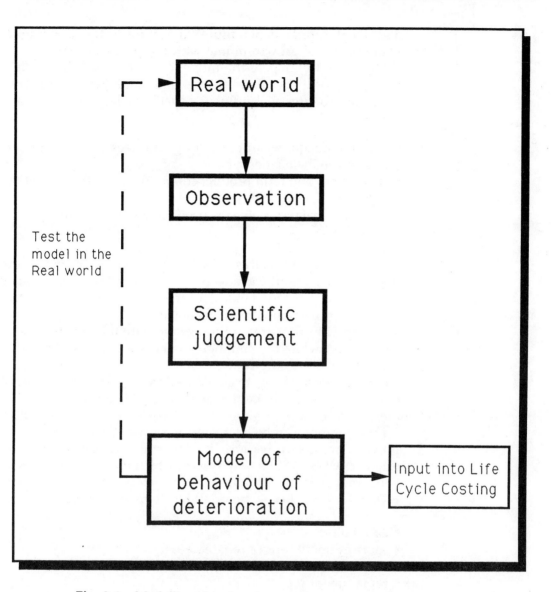

Fig. 2.4 Modelling deterioration.

properties. As buildings grow older they do not perform as well financially; their rates of rental growth slow down in comparison with trends in full market values. Investors need to plan for obsolescence either when calculating future returns on property assets because of petering rental growth, or through additional capital expenditure to refurbish the building to maintain rental growth in line with the property market.

Measures to mitigate the influence of obsolescence fall into two categories: specific measures and contingency measures. Specific measures will influence system selection and design and include:

- securing adequate support and guarantees by the suppliers during the installation phase;
- ensuring a long term product commitment on the part of the supplier;
- striking a balance between product innovation and proven reliability.

Contingency measures which make provision for future requirements might include provision of additional capacity which could be needed in the future and designing the system or building for modular expansion.

The specific measures are concerned with the current state of technology, while the contingency measures attempt to make predictions about the future, such as where we are going to and when we are likely to get there. In view of the time value of money, the second prediction is of equal importance to the first.

Deterioration can be overcome at a price. Obsolescence may prove more costly because it may entail replacement of an existing system with a new system which may be incompatible with the existing structure. Both deterioration and obsolescence will involve some or all of the following costs:

- adaptation
- refurbishment/retrofit/rehabilitation
- modernisation
- replacement/repair
- fitting out
- minor works to comply with legislation.

Many of these cost centres are used interchangeably by the building industry, sometimes being absorbed in the general heading of maintenance. For consistency, four headings are adopted:

- *refurbishment* which includes adaptation, retrofitting, up-grading and rehabilitation;
- *modernisation* involving minor works to modernise and improve a system or a property, much of the work being to mechanical and electrical engineering services;
- *maintenance* which includes replacing and repairing work, and minor works to comply with legislation, and
- *fitting out* is usually a capital cost which involves adapting and finishing a building or system to meet a tenant's needs.

The boundaries between the groupings are arbitrary, but the expenditure does need to be separated in some way, otherwise the hidden costs of obsolescence will be concealed in the maintenance expenditure. Furthermore, the maintenance budget will often conceal the upgrading of premises. Refurbishment includes the process of bringing buildings into line with present day needs. There can be hidden and expensive difficulties with refurbishment, because any building being substantially opened up must be brought into line with the current Building Regulations.

Flexibility and adaptability of the layout of a building should not be ignored at the design stage because refurbishment cycles are becoming shorter. A survey conducted by CALUS in 1986 asked a number of property investors, 'How long from date of initial construction do you think it will be before a landlord is involved in major refurbishment expenditure involving at least 30% of the cost of a new building?' To avoid a uniform response of 'every 25 years', investors were asked to assume that leases were granted for period of five years only on a full repairing and insuring basis. The response was as shown in Table 2.7.

Bearing in mind that the above figures relate to refurbishment, not to modernisation which is likely to be undertaken before the major refurbishment, they show that the expectations of building life cycles are becoming shorter. The figures also mask factors which will affect the perception of useful life. For instance, in high-technology office buildings there is likely to be a substantial proportion of mechanical and electrical services which are subject to rapid technological change and modernisation. In the case of warehouses, the pressure to produce inexpensive accommodation coupled with a high level of wear

Table 2.7 (*Source*: Depreciation of Commercial Property, CALUS 1986.)

Building Type	Number of years to refurbishment		
	Mean	Standard Deviation	Mode
High tech office	16	6	15
Office (single tenant)	19	6	20
Traditional warehouse	21	9	20
Retail warehouse	25	16	20
Shopping centre (mall)	11	4	10
Office (multi-tenant entrance hall)	11	4	10

and tear means that short life span and low capital cost equipment are often used; the result is a high maintenance cost.

Depreciation

Depreciation is the economic consequence of deterioration and obsolescence and Fig. 2.5 shows their relationship.

Understanding and allowances for the impact of depreciation on property should be included within the life cycle costing framework. Depreciation is mainly the concern of the private sector, rather than the public-sector organisations. That does not mean that the public sector should totally ignore depreciation, however, because all buildings have a value in the market place, and that value is affected by depreciation of property.

Traditionally, investment in commercial property has given long-term capital appreciation and also rental income. However, the differences in value between a new building and one 20 years old are such that in many locations the rental value of an unrefurbished 20-year-old building is no more than 60% of that of a new one and the capital value may be only 50% of its modern equivalent.

The treatment of depreciation of assets in company accounts

is standardised by the Accounting Standards Committee. Depreciation is defined by the committee as 'the measure of the wearing out, consumption, or other reduction in the useful economic life of a fixed asset whether arising from use, effluxion of time or obsolescence through technological or market changes'. This is rather a grand way of saying that things wear out over time and that this must be taken into account in a company's balance sheet, and the amount of depreciation disclosed in the profit and loss account. Depreciation is too important to be ignored. Buildings do wear out and the capital is thereby eroded. Methods of depreciating the value of buildings are complicated and their detailed discussion, other than broad principles, is not appropriate here.

Difficulties arise because property (land and buildings) is normally valued as a single entity, whereas depreciation

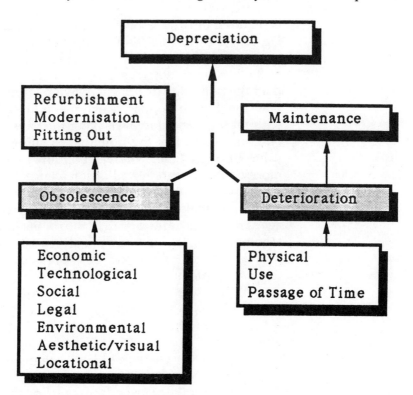

Fig. 2.5   Depreciation, deterioration and obsolescence.

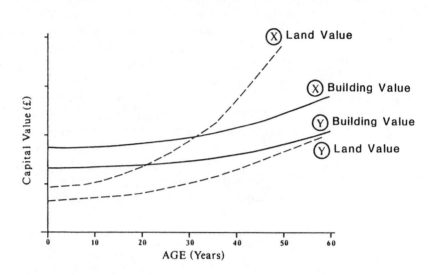

Fig. 2.6 The relationship between property value and land value.

is required only in respect of the building and components. As Fig. 2.6 shows, the relationship between land value and the property value is a function of many factors beyond the control of the building owner. In case X, the land value is worth more than the building after 25 years, whereas in case Y it is possible that the building may reach the end of its functional life without the site value overtaking the value of the unimproved building. Land has the potential to show constant growth in value through time, whereas buildings will be affected much more by deterioration and obsolescence.

It is possible to predict the approximate point in time when land value will overtake the value of a property increasing in age. The future annual growth rate in the land value is forecast, together with the future annual growth in the value of the property. The land value is expressed as a percentage of the total value of a new property:

$$(1 + G_A)^n = P_L \times (1 + G_L)^n \tag{17}$$

where
n = number of years until land value overtakes value of the property;
G_A = annual rate of growth in value of the property (if growth is 4.5% = 0.045);

G_L = annual rate of growth in value of land;
P_L = site value expressed as a percentage of the total value of the property (0.50 = 50%).

The depreciation of a building in the UK cannot be treated as a cost which can be set against earnings for tax purposes. The UK tax system is founded upon a rigid distinction between capital and revenue expenditure. Corporation tax is a levy on income and is not concerned with capital gains or capital expenditure. However, capital allowances are possible on certain items of plant and machinery.

It is easy to be confused about the way that depreciation is being shown in a company's accounts. There are three types of account that are prepared by trading companies:

- The *profit and loss account* shows the trading position for one financial year. The amount of depreciation on the fixed assets is included in the calculation of the profit or loss.
- The *balance sheet* shows the assets and liabilities of the company at a specific day. Fixed assets are depreciated to take account of their reduced value.
- The *tax equalisation account* is the gross profit less capital allowances and writing down allowances (depreciation) which give the profit subject to corporation tax.

The amount of depreciation provided in the balance sheet is not allowable in the company's calculation for tax liability. The main reason for this is that although a company in reporting to its shareholders is interested in calculating profits as accurately as possible, the Government is more interested in regulating investment. The underlying philosophy is that the Government wants companies to invest in plant and equipment to improve the nation's economic performance.

Life cycle cost planning does not specifically address the question of depreciation when calculating a net present value or annual equivalent value, because depreciation is an accounting exercise. However, the depreciation issues should be considered in the decision-making process.

Fiscal policy
The running costs of buildings such as energy, cleaning, maintenance, rent, general rates and insurances are allowable

as a business expense. At first sight this policy appears to support the low capital cost with high maintenance cost approach to building, because a given volume of maintenance work must be less expensive to the property owner than a corresponding volume of new work. In reality, such a short-sighted approach is not a sound basis for making investment decisions.

Salvage and residual values
The salvage value is the scrap value of a component. The residual value is much more problematical to quantify. For a complete building it is generally assumed that the residual value will equate to the value of the land with any site services provided. The logic behind this is that the building element of a property will be of nil value at the end of its useful economic life; but it must be remembered that the period of analysis may not coincide with the 'useful economic life'. What does this term mean? Definitions are hard to find, but the Accounting Standards Committee defines it as being, 'the period for which the present owner will derive economic benefits from its use'. A counter view is to use the market value for second-hand properties.

In the case of components or elements of a building the residual value is easier to quantify.

Forecasts and assumptions

Life cycle costing techniques are heavily dependent upon forecasts about the future, whether about the expected lives of components and finishes or about future maintenance expenditure. Some of the forecasts will be no more than expert judgement, best guesses, or hunches. Others will involve the use of forecasting methodology. Figure 2.7 shows the sequence of forecasting.

A vital element in forecasting is the ability to understand trends combined with the ability to visualise what could happen in the future. This calls for the exercise of imagination and insight in assembling from the mass of available infor-mation a vision of the direction future developments will take; for instance, what will happen to energy prices in the future. Forecasting relies upon the assumption that future costs can be

Fig. 2.7 The forecasting process.

predicted to some extent by referring to patterns of cost that have existed in the past. Such assumptions only hold true if no outstanding technological leap occurs and the existing stimuli remain constant both in type and degree (i.e. the push effect is unchanging). The further forward the forecast, the greater the likelihood of prediction errors.

The important part played by professional judgement should not be underrated. Sound judgement is required at all stages; in the selection of assumptions, determining and selecting the data required and choosing the most appropriate forecasting techniques to use. Analysts with biased assumptions may reach biased conclusions. The fault, however, lies not in the technique but in the analyst. It is people rather than analytical techniques, who make decisions.

Assumptions have to be made when any quantitative or qualitative information is either missing or unreliable, and these must be based upon some logical foundation. When dealing with buildings, long time horizons are involved and the analyst should always ask three questions:

- what should happen?
- what could happen?
- what does happen?

Most assumptions should be tested by the use of sensitivity analysis, which is discussed later.

Prediction and forecasting errors
The process of predicting the various costs and savings is fraught with errors, the root cause of which is uncertainty about the future. Prediction errors can be classified broadly as

measurement and sampling errors. Errors of measurement can occur as a result of differences in measurement units, while sampling errors result from the fact that a sample may not be representative of its population.

Prediction errors may occur as a result of incorrect assumptions about discount rates, inflation rates, lives of buildings and components, period of analysis, energy inflation rates, and allowances for maintenance. They will occur; the most important thing is to recognise where they are likely to occur and the likely magnitude of the error.

Calculating the effect of inflation

If inflation rates are expected to vary from year to year, different discount rates can be used for different years in an NPV calculation. Such a procedure, however, entails cumbersome calculations. If a nominal discount rate is used, the simple approach is to use a single average inflation rate with a single nominal discount rate. A preferable procedure is to use deflated cost streams with real discount rates which take inflation into account.

The real difficulty is the practical one of formulating explicit forecasts of future interest rates. This may explain why, in practice, both public and private-sector decision-makers work in terms of a 'long-term' interest rate which is assumed to be constant over time.

What happens when there is a differential inflation rate, which is the expected percentage difference between the rate of inflation assumed for a given cost item such as energy, and the general rate of price inflation? When the general rate of inflation moves at a different inflation rate for the time under consideration, a differential inflation rate adjustment must be made.

Cash flows

The time of occurrence of cash flows must be known in order to convert pounds spent or received in future years to equivalent values at a base time.

Cash flows may be single lump sum events, such as the payment for a repair or replacement, or they may be recurring in nature, such as planned, periodic or cyclical maintenance costs. Also, they may occur irregularly in both time and amount, such as emergency repairs. Cash flows may be spread out over the year, such as revenue from the monthly rental of a property. Generally the cash flows will be irregular, with some concentrated at certain times of the year and others spread throughout the year.

A simplified approach is usually adopted, rather than an attempt to model all the cash flows on a project. A common assumption is that cash flows occur either at:

• the beginning of the year being considered
• the end of the year, or
• the mid-point of the year.

Another option is to calculate the total cash flow and assume a continuous equal stream of money throughout the year. Obviously none of these options describes the exact pattern of actual cash flows, but in most cases absolute accuracy is not necessary for life cycle exercises.

Cash flows involve both costs and revenues, and they are affected by inflation; probably it will affect different components of the cash flows in different ways. Proper interpretation of any life cycle appraisal requires an understanding of how inflation has been handled in the cash flows.

Interest changes

A question of increasing concern to both public- and private-sector clients is how interest charges (financing charges) on borrowed capital should be accounted in a life cycle costing exercise. The examples presented thus far in this book and, indeed, in almost all life cycle cost applications tend to neglect this issue. Such an approach is valid so long as the initial capital cost is expended as a lump sum, but if this cost is financed through borrowing, explicit account may need to be taken of the resultant interest payments on the borrowing.

At the same time there is a danger of double counting because the process of discounting implicitly takes account of capital repayment. Therefore, care must be exercised in the incorporation of interest charges in the life cycle cost calculations.

The most effective approach is to detail *all* cash flows throughout the life of the project. A short-cut is available, as will be seen below, but the danger in adopting such a short-cut is either that certain costs will not be included, or that some costs will be counted twice.

There are two aspects to handling interest charges:

- when calculating the net present value in a life cycle costing exercise when more than one option is being examined;
- when calculating the running costs of a facility at today's prices; in this instance the sums of money paid or received in the future are not discounted.

To see how the net present value approach works, consider a simple example. It is proposed to finance a particular project with initial cost C_0 and period of analysis N years by borrowing. The interest charge is $i\%$ and the loan will be repaid in full at the end of the project. A real discount rate of $r\%$ is being applied to the project. The cash flows relating solely to the financing of the project over the period of analysis of the project can be detailed as follows:

Time period period of analysis	0	1	2	3	...	N
Money outflow	$C_0^{(1)}$	$iC_0^{(2)}$	iC_0	iC_0	...	$iC_0 + C_0^{(3)}$
less						
Money inflow	$C_0^{(4)}$	0	0	0	...	0
Net outflow	0	iC_0	iC_0	iC_0	...	$iC_0 + C_0$

$^{(1)}$ = capital payment paid to contractor
$^{(2)}$ = interest payments
$^{(3)}$ = capital repayment paid to financier
$^{(4)}$ = capital borrowing from financier

The present value of the net cash outflows is:

$$\text{LCC} = \frac{iC_0}{1+r} + \frac{iC_0}{(1+r)^2} + \ldots + \frac{iC_0}{(1+r)^N} + \frac{C_0}{(1+r)^N} \tag{18}$$

The interest payments represent an annuity of iC_0 extending over N years. Therefore, the annuity formula which gives the net present value of a constant annual sum can be used to simplify the calculations. An annuity of £1 at r% over N years has present value.

$$a_{rN} = \frac{[1 - 1/(1+r)^N]}{r} \tag{19}$$

Thus the present value of the interest payments are

$$\text{PV}_i = iC_0 \frac{[1 - 1/(1+r)^N]}{r} \tag{20}$$

The capital repayment at the end of the project has present value:

$$\text{PV}_c = C_0/(1+r)^N \tag{21}$$

Thus the life cycle costs of capital and interest charges in financing the project through this type of borrowing are:

$$\text{LCC} = \text{PV}_i + \text{PV}_c = iC_0 \frac{[1 - 1/(1+r)^N]}{r} + \frac{C_0}{(1+r)^N} \tag{22}$$

Now consider the, perhaps special, case in which the interest rate on the borrowed capital and the discount rate are identical. Then equation (22) simplifies as follows:

$$\text{LCC} = r\,C_0 \frac{[1 - 1/(1+r)^N]}{r} + \frac{C_0}{(1+r)^N} \tag{22a}$$

Cancel r in the first expression:

$$= C_0\left[1 - \frac{1}{(1+r)^N}\right] + \frac{C_0}{(1+r)^N}$$

Multiply through the bracket

$$= C_0 - \frac{C_0}{(1+r)^N} + \frac{C_0}{(1+r)^N}$$

$$= C_0$$

In other words, *if the interest rate and the discount rate are identical, interest charges can be ignored in the life cycle cost calculations.* They are, of course, still important in assessing the running costs of the project if the organisation faces a revenue expenditure constraint.

Equation (22a) can be used to simplify equation (22). Note that:

$$i = (i - r) + r \tag{23}$$

Thus equation (22a) can be rewritten:

$$LCC = (i - r)\, C_0 \frac{1 - 1/(1+r)^N}{r} + r\, C_0 \frac{1 - 1/(1+r)^N}{r}$$

$$+ \frac{C_0}{(1+r)^N} \tag{22b}$$

which from (22a) is:

$$LCC = (i - r)\, C_0 \frac{1 - 1/(1+r)^N}{r} + C_0 \tag{24}$$

In other words, when interest rate and discount rate are not identical, the life cycle cost implications of the capital plus interest charges can be calculated as follows:

(1) Calculate the value of an annuity of $(i - r)C_0$ at interest rate r over N years. (i = interest rate on the borrowed capital, r = discount rate.)
(2) Add the result of (1) to the initial capital costs.

Note that if $i > r$ interest charges increase life cycle costs, while if $i < r$ they decrease life cycle costs.

One final point is worth noting. The simplest case in which the loan is repaid in full at the end of the period has been taken. It is, of course, possible to use a mortgage with an annual repayment. The same kinds of calculations should be made. All that will change is that the final payment of C_0 will be replaced by an annual stream of capital repayments, and the annual cash flows will consist of interest charges plus capital repayments; the balance between these will, of course, change over the life of the product. *It remains the case that interest charges can be ignored if interest rate and discount rate are identical.*

Inflation again

We have indicated that the preferred method for 'dealing with' inflation is to use an inflation-free discount rate (the real discount rate) and express all into a constant (today's) price. This works so long as all costs are expected to inflate at the same rate. There is a minor complication if the aggregate inflation rate is expected to vary. Evidence supports the proposition, however, that there is a relatively stable relationship between 'the' inflation rate and market interest rates. Given this proposition, the real (inflation free) discount rate can be expected to be stable over time.

What happens, however, if different cost elements are expected to inflate at different rates, e.g. if labour costs are expected to inflate at a lower rate than energy costs? Two eventually equivalent approaches can be adopted to deal with this case:

(1) Perform the present value calculations in nominal terms: by expressing all prices and costs as current prices and costs and discount rates. Assume, for example, that annual costs consist of two cost streams C^1 and C^2. Expected inflation for cost stream C^1 (perhaps labour costs) is *1%* and for cost

stream C^2 (perhaps material costs) is $m\%$. Initial capital costs are C_0 and the market interest rate is $d\%$. Then the NPV is:

$$NPV = C_0 + \frac{C^1(1+l)}{1+d} + \frac{C^1(1+l)^2}{(1+d)^2} + \ldots + \frac{C^1(1+l)^t}{(1+d)^t} + \ldots \frac{C^1(1+l)^N}{(1+d)^N}$$

$$+ \frac{C^2(1+m)}{(1+d)} + \frac{C^2(1+m)^2}{(1+d)^2} + \ldots + \frac{C^2(1+m)^t}{(1+d)^t} + \ldots \frac{C^2(1+m)^N}{(1+d)^N}$$

Hence

$$NPV = C^0 + \Sigma_{t=0}^{N} \frac{C^1(1+l)^t + C^2(1+m)^t}{(1+d)^t} \tag{25}$$

(2) Choose one cost stream (the cost stream with the lowest expected inflation rate) as base. Assume that the inflation rate for that cost stream is $b\%$. Then adjust the market interest rate and all the expected inflation rates for all other cost streams by this base inflation rate using equation (9). For example, the adjusted discount rate is:

$$d^1 = \frac{1+d}{1+b} - 1 \tag{26}$$

and if the inflation rates for cost stream i is $i\%$ the adjusted inflation rate for cost stream i is:

$$i^1 = \frac{1+i}{1+b} - 1 \tag{27}$$

The NPV is then, for the case in equation (25)

$$NPV = C_0 + \Sigma_{t=0}^{N} \frac{C^1 + C^2(1+m^1)^t}{(1+d^1)^t} \tag{28}$$

In other words, a *differential* inflation rate given by equation (27) has been applied to cost stream 2 and the cost stream so generated is discounted at the adjusted discount rate [adjusted by equation (26)].

To illustrate this, consider a very simple example. Assume a 5-year shop fitting project with initial capital costs £10000. Annual maintenance costs are estimated at today's prices at £1000 for materials and £1500 for labour. All costs are assumed to be incurred at the end of the year. Material costs are estimated to inflate at 5% per annum and labour costs at 10%. The market discount rate is 15%.

Method (1)

$$\text{NPV} = 1000 + \frac{1000 \ (1.05) + 1500 \ (1.10)}{(1.15)}$$

$$+ \frac{1000 \ (1.05)^2 + 1500 \ (1.10)^2}{(1.15)^2}$$

$$+ \frac{1000 \ (1.05)^3 + 1500 \ (1.10)^3}{(1.15)^3}$$

$$+ \frac{1000 \ (1.05)^4 + 1500 \ (1.10)^4}{(1.15)^4}$$

$$+ \frac{1000 \ (1.05)^5 + 1500 \ (1.10)^5}{(1.15)^5}$$

$$= £10414$$

Method (2)
Take materials costs at base. The differential rate of inflation of labour costs is:

$$l^1 = \frac{1.10}{1.05} - 1 = 4.76\%$$

The adjusted discount rate is:

$$i^1 = \frac{1.15}{1.05} - 1 = 9.52\%$$

$$NPV = 1000 + \frac{1000 + 1500\ (1.0476)}{(1.0952)}$$

$$+ \frac{1000 + 1500\ (1.0476)^2}{(1.0952)^2}$$

$$+ \frac{1000 + 1500\ (1.0476)^3}{(1.0952)^3}$$

$$+ \frac{1000 + 1500\ (1.0476)^4}{(1.0952)^4}$$

$$+ \frac{1000 + 1500\ (1.0476)^5}{(1.0952)^5}$$

$$= £10414$$

Chapter 3 Types, uses, sources and output of data

The data required for a life cycle costing exercise will comprise a vast mass of (sometimes unrelated) information. It is important not only to understand the process from the initial data collection stage to the final output and decision-making point, but also to consider what types of data are needed and their sources; for example, buildings in use, or manufacturers' and suppliers' information. The usefulness of data should be evaluated in relation to their cost, and their means of storage should also be considered.

A major constraint and criticism on the application of life cycle costing techniques is the lack of an appropriate database on costs and performance. It is noteworthy that for new and secondhand cars there is an abundance of comparative data on models, price, running cost, and fixtures, yet for buildings, which represents major expenditure, very few organisations have any real idea of what the total running costs are of the buildings they own and operate. It is true that cars are production line products whereas buildings tend to be 'one-off' products. Whilst buildings may look different they perform a particular function and their costs should be managed and not merely monitored.

A turning point in the improvement of the collection of data will be the advent of more sophisticated computer assisted control systems and intelligent buildings.

Intelligent buildings are still in their infancy. There are different forms of technology serving an intelligent building but basically the building contains:

• Building automation systems that enable the building to respond to external factors and conditions (not just climatic, but also fire and security protection). The performance of the building is constantly monitored by computer and any necessary changes are initiated by the computer.

- Office automation systems and local area networks.
- Advanced telcommunications to enable rapid communication with the outside world via the central computer system.

The notion of intelligent building is increasingly linked with big business. There is dependence upon extensive, expensive information technology. The potential for technology obsolescence is high, but the running costs can be monitored closely. Whilst the inital capital costs may be higher than conventional buildings, the life cycle costs may be lower.

Types of data

The types of data required to carry out a life cycle approach to surface finishings fall within the following categories:

- cost data
- occupancy data
- physical data
- performance data
- quality data

The use of data

Most large organisations hold data in the form of fuel bills, maintenance records, water charges, and so on. Analysing such data on the running costs and performance of buildings in use is beset with difficulties. Any piece of information is a snapshot of a situation at a specific point in time. For instance, if the maintenance records of two university teaching buildings located on different sites were collected and analysed, the information would show the details of expenditure, the floor areas of the buildings and the maintenance undertaken over the past year. To enable a true comparison to be made, however, the buildings would need to be of the same age, the same construction type, and used for the same purpose. The figures are unlikely to reveal the hidden issues such as the useful life left in the components: for example, is major expenditure likely to be required to replace all the floor finishings in the next few years? Nor will it show the present, or any previous, policy on standards.

Furthermore, the data are unlikely to show the amount of maintenance that has been deferred because of a lack of funds. Too frequently the maintenance budget is allocated on the basis of what the organisation can afford, rather than what maintenance is desirable. Financial allocations are generally made on the basis of last year's expenditure rather than next year's need.

The period for which past data are both accurate and adequate must be carefully chosen. Should we look over the past year, five years or ten, and having obtained the data, what index should be used to update the information to the same point in time?

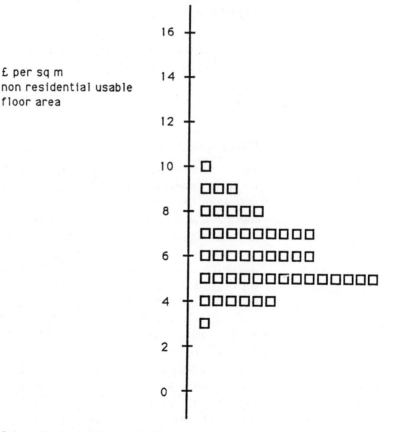

Fig. 3.1   University grants' committee survey.
(*Source*: University needs for maintenance and minor works, University Grants Committee.)

There are no universal answers, but trends in cost data can be identified and used for rebasing purposes. Building Maintenance Information Ltd (BMI) publishes various indexes that can be used to update the cost information to a common point in time. The data will be variable, as is illustrated in Fig. 3.1, which shows the results of a survey conducted by the University Grants Committee into the maintenance costs of non-residential university buildings.

The histogram shows the mode, but simple statistical calculations give both measures of central tendency and also an indication of the spread of the data.

Cost data

Data describe four dimensions: these are physical, performance, quality and cost. Cost data are most frequently available for materials and services or details of wages but without being supplemented by the other types of data, they are of uncertain value.

Occupancy data

They are used by the organisation's management and accounting processes or to comply with fiscal requirements. In many instances it can be an onerous task to convert them into feedback about running costs. For example, if an organisation uses directly employed labour to clean its premises, details of the wages bill and the materials purchased and used will be available, but no details are likely to be available about the difference in cleaning, for instance, carpeted areas as distinct from a quarry-tiled floor area.

Feedback is required about past occupancy running costs to enable assumptions that are implicit to be made explicit. Information about the use of the area being studied (function, level of occupancy and times of use) is frequently available in the form of cost and performance data, but, as noted above, may require some refinement. Table 3.1 shows occupancy cost data for some office buildings in Holland. Similar data are available for UK and US buildings. In the USA, the Building Owners and Managers Association (BOMA) collects data on expenditure for office buildings.

Table 3.1 Occupancy cost data for offices.
Data on occupancy costs from a premises survey conducted by William Jansen of the
Dutch Insurance Society, Centraal Beheer. (1986 price levels)
(Reproduced by permission from Facilities 5 (4), April 1987.)

Overheads/ sq m gross	145.22	318.42	192.84	240.48	240.00
Energy costs/sq m gross	18.70	22.60	13.38	21.33	24.00
Maintenance costs/sq m gross	4.34	44.60	5.40	4.19	6.00
Cleaning costs/sq m gross	13.04	21.53	12.93	25.80	13.00
Occupancy costs/sq m (gilders)	181.30	407.15	224.55	291.80	283.00
Date of building	1982	not known	1974	1973	1984
Gross building area sq m	2,300	2,390	7,500	9,930	5,000
Occupancy costs/employees (gilders)	5,213	10,136	9,624	6,900	7,294
Overheads/ sq m gross	240.61	295.16	344.80	254.75	161.58
Energy costs/sq m gross	91.52	31.41	30.28	19.04	23.99
Maintenance costs/sq m gross	22.14	22.63	28.67	16.53	70.35
Cleaning costs/sq m gross	21.60	24.39	14.91	37.80	29.18
Occupancy costs/sq m (gilders)	375.87	373.59	418.66	328.13	285.11
Date of building	1969	1977	1976	1975	1930/1959
Gross building area sq m	13,551	21,079	73,670	45,000	27,989
Occupancy costs/employees (gilders)	9,795	6,300	12,874	12,874	9,388

Physical data

There are data about the physical aspects of a building that can
be measured, such as the gross superficial floor area, the areas
and types of floor and wall finishings, and the number of
occupants using the building. With general accounting systems
these data are often entirely absent, but many organisations are
now developing property databases for storing these kinds of
information.

Performance data

Performance data relate in particular to the specification of the
surface finishings. They include the maintenance, replacement

and cleaning cycles of each finish together with other performance-based data such as the thermal conductivity of the material. Performance data would also include the times and level of occupancy within each functional space.

The way the areas are used is important. For instance, undoubtedly there will be a high frequency of use on the floor finishings to a shopping mall where there are large numbers of pedestrians passing throughout the day. There is also the need to maintain a high level of finish and this will dictate the type of maintenance which is undertaken. Equally, within the mall the service areas will have similarly heavy traffic, but there will not be the same demand for high standards.

Quality data

The data on quality are intended to give an indication of the finishings, but more importantly, in existing buildings are closely linked to a condition survey of a facility. This will identify the condition of the finishings at any given time, the maintenance expenditure necessary to bring the finishings to a required standard, and the estimated remaining service life. Such data may be used for general budgetary purposes and may also inform management of likely risks of failure in particular cases.

Quality data are influenced by policy decisions such as how warm the building should be, how clean it should be, and how well it should be maintained. Data relating to quality are highly subjective and less readily accountable and trustworthy than cost or performance data which can be objectively measured. However, that is not to say that the analyses are not worth doing; it simply means that the level of professional skill and judgement used has to be that much higher.

Data sources

Data for life cycle cost purposes should be taken from three main sources:

- *Specialist manufacturers, suppliers and specialist trade contractors and government agencies.* The manufacturers and/or suppliers of materials and components may be expected to know not only the cost of their products, but their anticipated lifespan and their maintenance and cleaning requirements.

- *Predictive calculations.* The simplest approach is to develop models which, say for cleaning costs, should take account of the superficial areas of all the surfaces to be cleaned, the frequency of cleaning, the cleaning standard required, the productivity of the cleaners, the types of areas and materials to be cleaned, and any special requirements. From this information unit price rates can be calculated upon which an annual cleaning cost can be based. This is a simple model but quite sophisticated predictive calculations can be made in this way, which reflect the likely level of certain running costs.

- *Historical data.* There are data from existing buildings in use. Some data are published, such as the BMI occupancy cost analyses, or may be found in technical journals. Other data may be obtained from clients' or surveyors' records.

Table 3.2 shows a study of average occupancy costs conducted by BMI. All costs have been adjusted to third quarter

Table 3.2 Occupancy cost analyses.
(*Source*: BMI, Average occupancy costs, December 1985.)

Elements	Building Type			
	Industrial buildings	Offices	Universities	Student hostels
Decoration	210	147	131	209
Fabric	127	209	131	155
Services	183	271	154	187
Cleaning	406	506	466	797
Utilities	1756	833	846	738
Administrative costs	416	1043	778	502
Overheads	890	1594	793	792
Total £/100 sq m pa	4103	4826	3293	3378

Table 3.3 Elements of property occupancy as defined by BMI.

Group element	Sub-element
Decoration	External decoration Internal decoration
Fabric	External walls Roofs Other structural items Fittings and fixtures Internal finishes
Services	Plumbing and internal drainage Heating and ventilating Lifts and escalators Electric power and lighting Other mechanical and electrical services
Cleaning	Windows External surfaces Internal
Utilities	Gas Electricity Fuel oil Solid fuels Water rates Effluents and drainage charges
Administrative costs	Services attendants Laundry Porterage Security Rubbish disposal Property management
Overheads	Property insurance Rates

1985 levels. The costs are given in accordance with a standard list of elements of property occupancy as defined by BMI and given in Table 3.3.

The BBA (British Board of Agrément) assesses components and systems for use in buildings. A certificate is issued which gives information on performance. This involves not only measuring certain performance characteristics of the product when new, but also predicting over what period of time it will maintain these characteristics. In the case of tiling, the BBA is helped by the fact that it is considering a single product of defined composition manufactured to an agreed quality specification with defined limits and installed in a structure in a specified manner. The predicted service life of tiles involves a study not only of the changes that will occur owing to environmental agencies, but also of changes produced by non-environmental agencies, such as fatigue loading, abrasion, and so on.

It is not only the British who have an Agrément scheme that considers service life. For example, the Centre for Better Living in Japan, which is running an Agrément approval system, publishes a list of expected service lives for components. Tables

```
BCIS ON-LINE - AVERAGE BUILDING PRICES
TYPE OF WORK : New build
ELEMENT : Ceiling finishes
SPECIFICATION - TABLE 45 CEILING FINISHES
             - Plasterboard


The rate is the cost excluding prelims divided by the
gross internal floor area

Based on an index of 309 at 3Q88 from series 0100
Sample size = 324 (excluding 2 rates with an answer of zero )
Mean = 10    Median = 8    Mode = 6
Range = 0 to 32    Standard deviation = 6
Quartiles: 1st = 6   3rd = 12
Deciles: 1        2        3        4        5        6        7        8        9
         3        5        6        7        8       10       11       13       17

        40 I
        36 I       O
        32 I       O
        28 I       O OO
        24 I       OOOOOOO
        20 I       OOOOOOOOO
        16 I O     OOOOOOOOOO
        12 I O OOOOOOOOOOOO
         8 IOOOOOOOOOOOOOOOOO O O
         4 IOOOOOOOOOOOOOOOOOOOOOOOOOOOOO O O O
         0 +-------------------------------------------------------
           ^          ^         ^         ^         ^         ^
           0         10        20        30        40        50
  Y-AXIS = Frequency
  X-AXIS = Cost (exc prelims.) / m2 gross floor area
```

Fig. 3.2 On-line facility. (Copyright RICS.)

of expected lives of various wall, ceiling and floor finishings are published by some agencies.

Initial capital cost data for finishings are available on the central Building Cost Information Service (BCIS) 'on-line' cost database which can be accessed quite simply using a micro-computer and modem unit. This form of data retrieval has the benefit that the data are easily and quickly accessible and can be rapidly analysed. The BCIS 'on-line' cost data are updated almost immediately as new cost information becomes available. As a result, the database represents a good reliable source of initial cost information.

Initial capital cost data for the installation of various different finishes can be obtained via the 'Average Building Prices' section of the database, which is one of the services available on the 'on-line' facility. This sub-divides each of the major cost elements. Cost data for individual finishes can be obtained from a sample of cost data for floor, wall or ceiling finishes. These cost data may then be presented graphically showing measures of central tendency (mean, median, mode) and of dispersion (range, quartiles, standard deviation). Figure 3.2 shows one example and illustrates the initial capital cost for a plasterboard ceiling finish.

Data output

The final output from the data modelling process should convey to the decision-maker the maximum amount of infor-mation in the simplest possible format. The life cycle cost profiles and sensitivity analyses, which will be developed in more detail in the following chapters, are particularly appropriate.

The life cycle cost profiles over a particular time horizon give a clear indication of the cost performance of each finish. Additionally, crossover points are identified where one finish becomes, in tangible life cycle cost terms, more or less economical than another.

Taking any example, it is possible, at a glance, to determine the NPV of any one of the finishes at any point in the time horizon or alternatively to measure the increase in life cycle cost between any two points in the time horizon. It is also possible to select a particular period of analysis and determine which finish has the lowest life cycle cost for that period of analysis.

Sensitivity analysis, to be discussed in Chapter 8, has the benefit of showing to the decision-maker which of those parameters within the exercise, such as life expectancy or discount rate, are the most critical in determining the NPV of that particular finish. This will identify those areas within the exercise where accuracy of initial data should be checked.

Chapter 4 Managing risk and uncertainty

The building industry is subject to more risk and uncertainty than perhaps any other industry. The production process is complex, generally unstandardised, and time consuming. It involves the co-ordination of a wide range of disparate yet interrelated activities and is subject to many external uncontrollable factors. Effective cost planning must take account of these risks and uncertainties if clients are to obtain buildings at an acceptable cost and contractors are to produce those buildings at an acceptable profit margin. Risk management has begun to be incorporated into building cost planning and life cycle costing, but remains a comparatively peripheral activity.

There are many reasons for the relatively conservative pace at which the building industry has introduced risk management to project appraisal. One has been the lack of practical systems tailored to the special needs and context of the industry. The theory of risk management is well understood, but it is only in recent years that it has begun to be converted into practical, microcomputer-based systems designed specifically for the industry. Various risk management systems and methods for 'dealing with' risk will be outlined in this chapter. One of these, sensitivity analysis, is considered in further detail in Chapter 8.

Life cycle costing, by definition, deals with the future, and the future is unknown. All cash flows associated with a particular choice of finish are only estimates, no matter how good the data on which they are based. Similarly factors such as the discount rate, replacement and maintenance cycles and the expected service lives of materials are also estimates.

None of these estimates can be made with certainty. At the same time, the decision-maker can be presumed to have some knowledge regarding the reliability of the estimates. Risk and uncertainty do not imply ignorance. Rather, they imply an ability to identify a well-defined range of possibilities with well-defined cost and environmental parameters.

Even if this range is confined to a simple three-way classification of 'optimistic', 'pessimistic', 'most likely', a complex basic description of an investment will result: with three factors such as initial capital cost, running cost and expected service life, potentially there would be $3 \times 3 \times 3 = 27$ possible NPVs to be calculated. It is necessary, therefore, that a simple management tool is developed capable of handling this complexity.

Expected value and adjusted discount rate

One suggested method for incorporating risk into investment appraisal is to calculate the *expected* NPV (ENPV) of the investment and choose on the basis of that expected NPV. The calculation of expected NPV makes use of the probabilities associated with the range of outcomes noted above.

To see how this works, consider the simple example in Table 4.1. Assume that a particular choice of finish has the optimistic, most likely and pessimistic cost flows as indicated, and that the estimated probabilities of these cost flows are also as indicated: note that these probabilities must sum to unity. The NPV of the optimistic, most likely and pessimistic cost flows are calculated respectively as $NPV_1 = £1929.71$,

Table 4.1   Expected net present value.

	Optimistic	Most Likely	Pessimistic
Initial Capital Cost (£)	1000	1500	2000
Outgoings(£) Year 1 2 Net Present Value (discount rate 5%)	500 500 1929.71	550 550 2522.68	650 700 3253.97
Probability Expected NPV (£)	0.2	0.6 2550.34	0.2

$NPV_2 = £2522.68$, $NPV_3 = £3253.97$. Expected NPV is then defined as:

$$ENPV = P_1.NPV_1 + P_2.NPV_2 + P_3.NPV_3 \qquad (29)$$
$$= 0.2 \times £1929.71 + 0.6 \times £2522.68 + 0.2 \times £3253.97$$
$$= £2550.34$$

Why should ENPV be used to guide decisions? The reason lies in the interpretation of ENPV. If the investment underlying the example of Table 4.1 were to recur, say 100 times, then the average NPV of these investments (the NPV of each investment summed over the 100 repetitions, and this sum divided by 100) would be very close to £2550.34. Thus ENPV represents a long-run average that would be achieved by this investment.

The example of Table 4.1 is considerably simplified. In particular, it assumes that risk applies to the stream of costs rather than the individual costs. If this assumption is dropped, the ENPV would have to be calculated by taking all possible combinations of cost. Then if the period of analysis is known only within a probability range, this factor would also have to be added; leading back, yet again, to the complex situation discussed above.

In such circumstances, ENPV does appear rather limited, as it produces a single figure. The probabilistic approach to be discussed below would be more appropriate in this more complicated, but perhaps more realistic, environment.

It has also been suggested by some analysts that risk can be handled by use of an appropriate risk-adjusted discount rate; that is, adding a risk premium to the standard discount rate discussed in Chapter 2:

$$d_0 = d_i + d_r$$

where

d_0 = risk adjusted discount rate
d_i = risk free discount rate
d_r = risk premium.

Levy and Sarnat (1977, p.193) show that adjusting the discount rate in this way is 'conceptually equivalent to the more sophisticated techniques based on the discounting of certainty–equivalent cash flows'. The problem lies first, in identifying an appropriate risk premium. Also calculation of the 'certainty–equivalent' cash flows involves calculating expected cash flows and so is subject to the same problems as those discussed with reference to the example in Table 4.1. It would appear, therefore, that a rather different approach is needed.

A risk management system

Using the terminology of Chapter 1, what is needed is a risk management system that relates to life cycle costing as in Fig. 4.1. Such a risk management system consists of three elements as shown in Fig. 4.2, each of which in turn consists of a series of sub-systems.

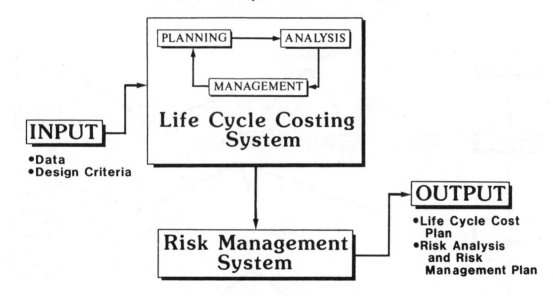

Fig. 4.1   Life cycle costing and risk management.

Fig. 4.2   Risk management system.

Risk identification

Every element of the building cost estimating process is subject to some degree of risk. These risks can be subdivided into those that affect quantities, unit price rates and the construction schedule. A decision tree approach is therefore appropriate as discussed by Chapman (1979) and Cooper *et al.* (1985), which subdivides the overall cost estimate into its major constituent parts, similar to the concept of levels developed in the work on life cycle costing by Flanagan and Norman (1983), see Fig. 4.3.

The appropriate level of disaggregation will vary across different building components and systems. For each baseline item in the decision tree the appropriate risks can then be identified. Note that some of these risks may be common to a number of baseline items (for example, general escalation, labour costs, quantities) and so will set up interdependencies between them.

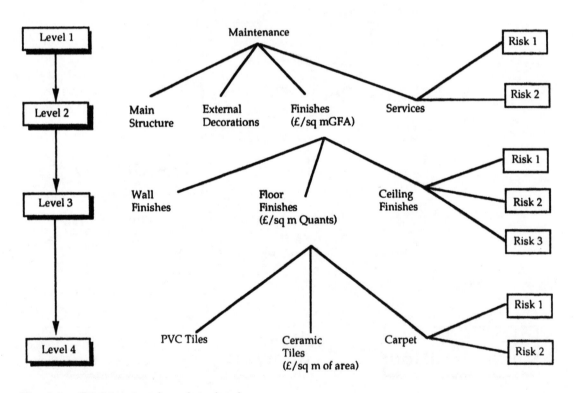

Fig. 4.3 Decision tree based on levels.

Risk analysis

Two approaches to risk analysis can be suggested: these are sensitivity analysis and the probabilistic approach/Monte Carlo simulation. The simple distinction between them is that sensitivity analysis does not require that a probability distribution be associated with the risk element. Alternatively, sensitivity analysis can be seen as a special case of a probabilistic approach, in which an equal probability is assigned to each value in the range over which the risky parameter is expected to vary.

In addition, sensitivity analysis identifies the impact of a change in a single parameter value within a project, whereas the probabilistic approach is a multivariate approach.

Sensitivity analysis
As has been indicated, this identifies the impact on NPV of a change in a single parameter used in the calculation of NPV; for instance, discount rate, initial capital cost and annual maintenance cost. There are many ways in which the results of a sensitivity analysis can be presented.

One such method, the spider diagram (see Perry and Hayes, 1985b) is considered here, and a related method is discussed in more detail in Chapters 8 and 9. The construction of a spider diagram such as that illustrated in Fig. 4.4 requires a series of steps:

Step 1 Calculate life cycle cost (LCC) for a proposed project in the normal way using best estimates of all parameters.

Step 2 Identify parameters subject to risk and uncertainty; for example, initial capital costs, maintenance cycles.

Step 3 Choose one parameter identified in Step 2 and recalculate LCC assuming that this parameter is varied by $\pm x\%$, where x lies in some predefined range; for example, recalculate LCC assuming that initial capital costs are changed by +1%, +2%, +... +10%, or −1%, −2% ... −10%.

Step 4 Plot the resulting LCCs on the spider diagram, interpolating between them to obtain intermediate values—for instance, the impact of an increase in initial capital costs of 1.2%. This generates the line labelled 'Parameter 1' in Fig. 4.4.

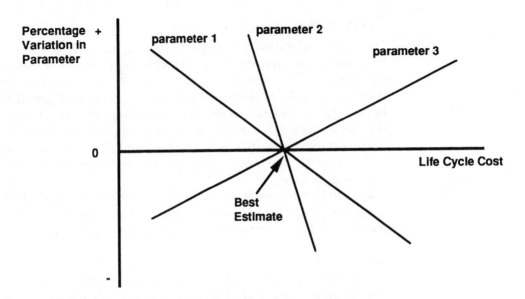

Fig. 4.4 Sensitivity analysis – the spider diagram.

Step 5 Repeat steps 3 and 4 for the remaining parameters in Step 2, to give the other lines in Fig. 4.4.

Each line in the spider diagram indicates the impact on life cycle cost of a defined percentage variation in a risky parameter. The flatter the line the more sensitive will life cycle costs be to variation in that parameter. This has two useful implications:

• it identifies those parameters on which attention should be concentrated in order to improve the certainty of the LCC estimate–there is little point in expending effort in improving the estimate of a parameter to which LCC is relatively insensitive;
• it emphasises to the decision-maker the fact that the LCC can be known only within some defined range.

One weakness of Fig. 4.4 is that it gives no indication of the likely range of variation of the risk parameters. This can be easily overcome, however. The decision-maker can be asked to identify the range within which a particular parameter is expected to lie, at a defined level of probability. For example, he might estimate that there is a 70% probability that maintenance costs will lie in the range (+8%, −6%) of the best estimate used in the initial analysis, and a 90% probability that the range is (+10%, −8%). This exercise can be repeated for each parameter to give the probability contours of Fig. 4.5.

It must be emphasised that probability contours are subjective estimates of the likely range of variation of life cycle costs. In addition, there is the underlying assumption that only one parameter at a time is being varied. Nevertheless, these contours provide valuable managerial information both on the robustness of a particular life cycle cost estimate, and on the parameters to which it is particularly sensitive.

There remains the question of how sensitivity analysis in the form of a spider diagram might be used to guide choice between different kinds of finishes. This is a 'state of the art' question that is unresolved: however, it is possible to give some flavour of the type of comparisons to be made, and the factors on which decisions might be based.

Assume that two finishes, A and B, are being compared.

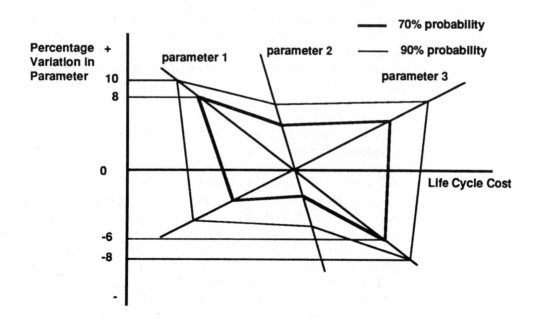

Fig. 4.5   Sensitivity analysis – probability contours.

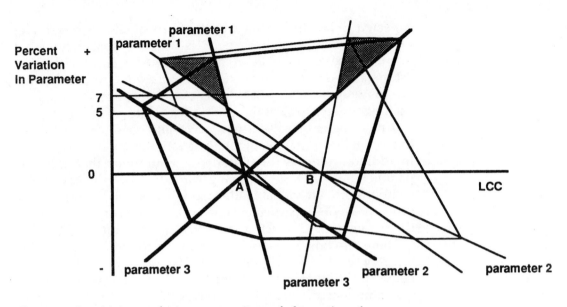

Fig. 4.6   Sensitivity analysis – comparison of alternatives 1.

Spider diagrams for each finish should be constructed and plotted on the same diagram. Figure 4.6 illustrates the resulting combined spider diagrams and the 70% probability contours for the finishes in a particular functional space. Best estimates of the various life cycle cost parameters result in finish A being preferred, with lower life cycle costs: but if parameter 1, which might be discount rate, were to be anything more than 5% above the currently-estimated value, finish B would be the lower-cost finish.

This type of comparison can be repeated for each parameter to indicate those for which the rank ordering of the two finishes would change: these are shown by the shaded areas in Fig. 4.6. In this example, finish B would be the lower-cost finish if the discount rate were 5% above its initially estimated value (if it were 5.25% rather than an expected 5%) or if parameter 3, which might be cleaning costs, were more than 7% above its initial estimate. By contrast, the rank ordering is unaffected by variation in parameter 2. The greater the extent to which the rank ordering would be changed by a parameter variation within the subjectively estimated probability contour, the less clear-cut is the advice to reject finish B in favour of finish A.

Figure 4.7 extends this analysis by introducing a further

dimension: the degree of sensitivity identified by the spider diagram. Again, an example is illustrated in which finish A has the lower life cycle costs on best estimates of the parameters, but finish A is also much more sensitive to these parameters. There would be a case for a client who does not like surprises to choose finish B: the higher cost might well be justified by the increased certainty of its not being exceeded to any great extent.

Some of these issues will be considered again in Chapter 8, where the analysis will also develop a somewhat different diagrammatic tool that is, in fact, closely related to the spider diagram described above.

This discussion shows that sensitivity analysis is no substitute for managerial decision-making. There is, as yet, no definite way of choosing between finishes A and B in Figs 4.6 and 4.7. Research on how such choices might be made continues: it is closely related to developments in social choice theory and choice under uncertainty. Nevertheless, sensitivity analysis even as currently developed is an essential component of managerial decision-making.

A major limitation of sensitivity analysis is that it is

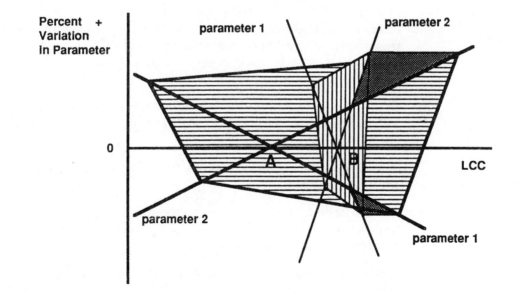

Fig. 4.7   Sensitivity analysis – comparison of alternatives 2.

univariate: only one parameter can be varied at a time. It is, however, quite clear that if there are risky parameters, several may vary simultaneously. In order to deal with this a multivariate method of risk analysis is needed: probability analysis.

Monte Carlo simulation

There are many ways in which probability analysis systems can be constructed, but the most easily used, yet extremely powerful, system is the Monte Carlo simulation. Derived originally from gambling models, Monte Carlo simulation is a means of examining certain types of problems for which unique solutions cannot be obtained. The method involves the introduction of random numbers, and so it does not yield a unique answer but serves to indicate the general range within which a solution may lie. This simulation system proceeds through a number of steps as illustrated in Fig. 4.8.

Step 1 Break the total project down into a number of basic items for analysis – perhaps by using the decision tree and levels approach discussed on p.74.

Step 2 Identify the risks associated with each of these basic costs and express each of them in the form of a probability distribution; for example, identify a probability distribution for cleaning costs, initial capital costs, or the expected service life of the finish.

Step 3 Use a random-number generator to select a random value from the probability distribution for each of the parameters in Step 2.

Step 4 Use these random values to calculate a life cycle cost, and store this cost.

Step 5 Repeat Steps 3 and 4 the required number of times – somewhere between 100 and 1000, depending upon the number of risky parameters.

Step 6 Plot the stored values in Step 4 as a probability distribution and cumulative distribution and calculate measures of central tendency and dispersion.

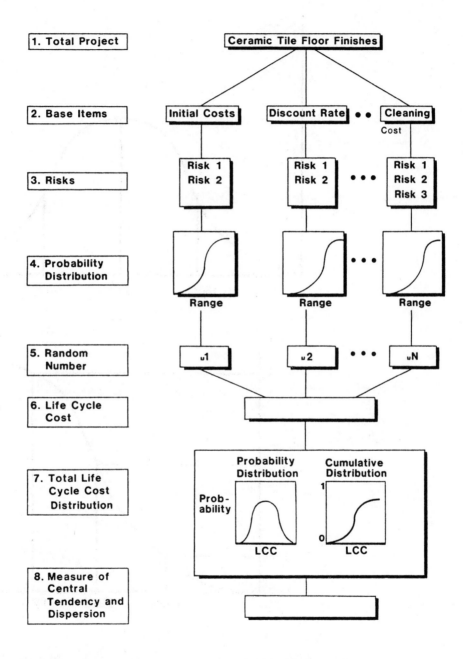

Fig. 4.8 Probability analysis – Monte Carlo simulation.

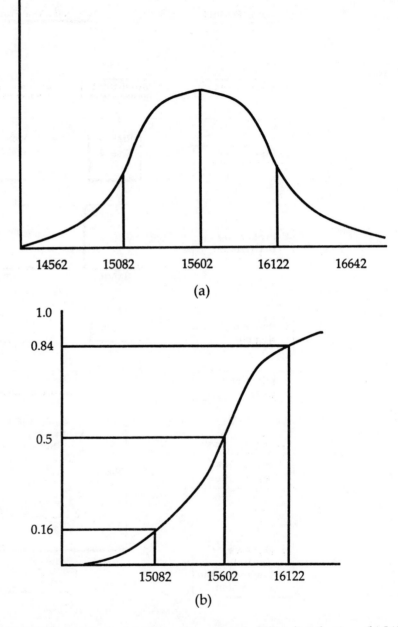

Fig. 4.9 Probability analysis. (a) Life cycle cost probability distribution. (b) Life cycle cost cumulative distribution.

The resulting probability and cumulative distributions will look something like those in Fig. 4.9. According to the central limit theorem, the overall probability distribution of life cycle costs will approximate to the normal distribution. These distributions have two uses. Firstly, the measure of central tendency – mean, median or mode – gives an indication of the most likely life cycle cost: if the probability distribution is approximately normal then mean, median and mode will be nearly identical.

Secondly, the measure of dispersion gives a measure of the confidence that can be placed in the 'most likely' value.

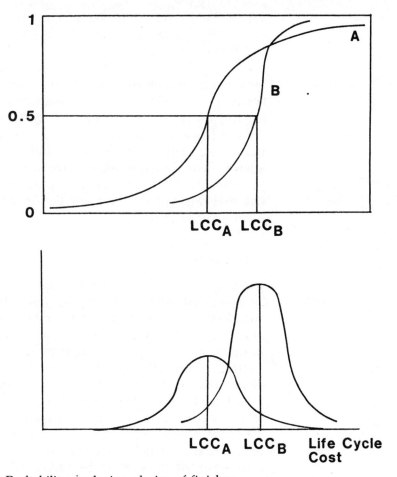

Fig. 4.10 Probability analysis – choice of finishes.

A powerful property of the normal distribution is that approximately 95% of all values lie within two standard deviations of the mean and 68% lie within one standard deviation. Thus if the probability analysis had generated the values: mean life cycle cost = £15602 and standard deviation = £520, there is a 95% probability that life cycle cost lies in the range £14562 to £16642, and a 68% probability that the life cycle cost lies in the range £15082 to £16122.

The cumulative distribution can also be used to indicate the probability that the life cycle cost will not exceed a particular value. For example, Fig. 4.9 indicates that there is a 97.5% probability that cost will be less than £16642 (since 2.5% of all values lie in the right hand tail more than two standard deviations above the mean) and an 84% probability that cost will be less than £16122.

Once again, consider how this technique will guide a choice between optional finishes. How might a choice be made between the optional finishes illustrated in Fig. 4.10? Finish A has the lower expected (mean) life cycle cost, but is much more risky: it has a much more dispersed probability distribution and flatter cumulative distribution. There is no simple, definitive answer. The decision-maker must weigh the implied trade-off between the lower expected costs of finish A and the higher risk that this cost will be exceeded, by an amount sufficient to justify choice of finish B.

Risk response

The final element in the risk management system of Fig. 4.2 can be described very briefly. It identifies the actions the analyst and decision-maker should take, given that they are operating in a risky environment. These can take two broad forms: risk transfer, for example by appropriate choice of construction contracts and management contracts; and risk control, by design choice and more detailed investigation of risk sources.

Risk transfer, while important, is generally outside the scope of the particular application being considered in this book. When applying life cycle costing techniques to a complete building, there will be extensive scope for choosing the form of contract most suitable to the identified risk exposure. This is less so when life cycle cost techniques are applied to individual building components. Nevertheless, it may be possible to obtain guarantees of performance or cost from particular suppliers, or to insure against failure. The risk analysis discussed previously will identify those situations in which such guarantees or insurance are most justified.

Risk control is of direct relevance both for complete buildings and individual building components. The discussion of sensitivity analysis and probability analysis systems indicates that it is possible to identify those parameters and environmental variables to which life cycle cost estimates are particularly sensitive. This can be used to guide attempts to improve the information base, and to reduce risk exposure.

An important point in this respect has been identified by Baker (1985, 1986). Additional information that reduces the degree of risk is valuable, but does incur costs for collection, storage and analysis. It should be obtained only if its value at least equals its cost. The risk management system should be capable of guiding such value/cost choices.

A simple illustration may make this clearer: returning to Fig. 4.6, there is little value in improving the estimate of parameter 2 if attention is being focused on the rank ordering of finishes A and B. On the other hand, it would be useful to obtain firmer information on parameter 1, the discount rate. The amount which may be spent in doing so can be estimated from a calculation of the cost of making the wrong decision.

The importance of judgement

Dealing with risk and uncertainty is an art that can be made to look like a science. We have described a series of methods for identifying the source, effects, and possible responses to risk and uncertainty but none of these techniques or systems remove the need for the analyst and decision-maker to apply judgement in the final choice of finish. There will be, inevitably, a degree of subjectivity in this judgement. What a risk management system does is to identify the areas in which judgement is necessary, the areas on which additional effort should be expended to reduce risk, and some of the likely consequences if a wrong decision is made.

Chapter 5 *Identification of costs and benefits*

This chapter discusses the importance of considering both costs and benefits in a life cycle cost exercise. The distinction between them is not always obvious, but failure to incorporate all relevant costs and benefits will certainly affect the validity of the exercise.

Chapter 2 introduced briefly the notion of calculating net cash flows as part of life cycle costing: net cash flows are simply revenues minus costs. Although most life cycle costing exercises follow the conventional line and concentrate on the cost side of the appraisal, it should be recognised that in many cases the benefits of each option must also be taken into account. How, then, can the problem of calculating net cash flows be approached? This is a two-stage exercise: firstly, the relevant costs and benefits to be included in the analysis must be identified; and secondly, these costs and benefits must be quantified.

Cost identification

Costs are of two basic types: tangible and intangible. As their name suggests, the former are usually readily quantified. Identifying the latter requires some imagination, and their measurement in monetary terms tends to be more problematical. Intangible costs are likely to be more relevant to public-sector projects than to those in the private sector.

Tangible costs

Tangible costs can usually be divided into two distinct categories: initial capital costs, and running costs.

In the case of internal finishes, the tangible initial capital costs include the outlay on materials, their installation costs and any associated construction costs. Tangible running costs, on the other hand, are the costs associated with activities such as the maintenance, replacement, and cleaning of the materials once they are in place.

An illustrative check list of relevant tangible capital and running costs for finishes would include:

Initial capital cost
• bedding
• finishes
• adhesives
• other allied construction costs.

Running costs
• cleaning
• fuel
• general rates
• insurances
• annual and recurrent maintenance
• replacement
• alteration/adaptation
• financing
• alternative temporary accommodation (decanting)
• disposal costs net of salvage or residual value.

The initial capital costs are largely self-explanatory. In addition to the materials themselves, it may be necessary to incur additional costs at the time of building or refurbishment related to preparation of surfaces, fixing of finishes (such as would be the case with floor tiles or wallpaper) and making good. The common element which links all the costs within this group is that they are one-off costs incurred at the beginning of the life of the project which is being appraised.

In contrast, the common element linking running costs is the fact that they are costs incurred throughout the life of the project in question. Once again, most of the costs included under this heading in the illustrative checklist are obvious, but one or two may benefit from some explanation. General rates are included because, to the extent that different finishes affect the assessed rental value of a project, the choice of finish will affect the rate bill received from the local authority. Similarly, differences in the technical specifications of finishes, particularly in relation to fire retardation, may affect the insurance premium to be paid on the finished project. The financing costs included in the list refer to the interest which must be paid on any funds which were borrowed in order to install a particular finish in the first place. To the extent that different amounts have to be borrowed to provide different finishes, the future stream of debt service costs will vary, and must be taken into account in the analysis.

Finally, 'decanting' costs are the costs of providing alternative accommodation while a particular finish is being installed during a refurbishment operation. If different finishes require either more or less extensive re-accommodation, or variable periods of time, then it is necessary to include these differential costs in the analysis.

As emphasised earlier, the choice of an appropriate period of analysis is important, especially when considering the running costs and replacement costs of each option. Running costs continue to be incurred with the passage of time. However, the magnitude of these costs will probably differ from one finish to another, so that if too short a period of analysis is chosen for comparison, a client could find himself locked into an option which is ultimately more expensive. As far as replacement costs are concerned, too short a period of analysis may result in the life cycle cost of one or more finishes being calculated at an artificially low level because their replacement at some stage during the life of the building has not been taken into account.

Such a schedule of tangible capital and running costs must be drawn up for each option and the costs converted into net present value terms for comparison. As far as tangible costs are concerned, this is a relatively simple exercise involving the techniques discussed in Chapter 2.

Intangible costs

Intangible costs should feature in the decision process because, although they may be difficult to measure directly, in some cases they have a decisive role to play. Take the case of a private-sector client, such as a shop owner deciding upon an internal finish in a food hall. If the customers are antagonistic to the finish chosen because they do not consider it to be hygienic, they will take their custom and goodwill elsewhere.

This loss of business is a cost and should be taken into account in the calculation.

Such an example also provides a good illustration of one of the common pitfalls of this kind of analysis. The loss of business has been identified here as a relevant cost to be included in the appraisal. If benefits are also being considered care should be taken to ensure that double-counting does not take place; in this example if the loss of business were also reflected in a lowering of the anticipated future stream of revenues flowing from the project. The loss of business can be considered either as an increase in costs or as a reduction in revenues, but not as both. Disruption costs during installation or renovation are another intangible cost. They can result in a loss of productivity, custom and profits. Once again some effort may be required to identify and quantify these costs.

So how are intangible costs measured? A great deal comes down to informed judgement based upon the experiences of those who have an intimate working knowledge of the situation. How should intangible costs be treated? Should they be incorporated into the costings along with the tangible costs? No: given the uncertainty of their scale, it is best to consider intangible costs separately. Indeed the most appropriate way of including them is to ask 'how big would these intangible costs have to be before they would influence the final decision?' This is the essence of sensitivity analysis. If the analysis shows that the margin of error on intangible costs is much greater than that which could be reasonably supposed to be associated with the measurement of these costs, they will offer no problem to the decision-maker. On the other hand, if they fall within the expected margin of error they will play a more strategic role and must be examined more carefully. In such cases the appraisal cannot give an unambiguous answer but can high-light the important cost areas which must be considered in a more detailed manner.

Benefit appraisal

The first step in identifying an appropriate measure of the benefits of a certain choice is to use the set of objectives laid down at the first stage of the appraisal. Systematic benefit analysis continues with the devising of measures which indicate the degree to which each option meets the stated selection criteria for the finishes. These are discussed in detail in Chapter 6.

As with costs, benefits may be tangible or intangible. Because many of the benefits are mirror-images of those items which have already appeared in the list of costs, unless the two listings are compared carefully there may be an element of double counting. For example, ease of maintenance should have been taken care of in the quantification of maintenance costs; it should not also appear as an intangible benefit. If double counting occurs, an option might seem to be less costly than it really is. This bias must distort the decision being made. Also, as with costs, placing monetary values on each item of benefit may not always be easy. When someone decides to purchase an internal finish he is buying a package or a bundle of benefits. He might not know how much he values each benefit individually but he does know how much he is prepared to pay for the total package.

How can the decision-maker carrying out the life cycle costing exercise find out how much the client is prepared to pay for the package? Often, it is impossible, at least directly. It may be that professional judgement leads to a knowledge of how much the client can afford. However, until the client is faced with a series of costed options he is often in no position to know whether or not he is prepared to pay.

Providing the client with a list of benefits forces him to think carefully about the package being purchased. As those who market products are only too well aware, consumers are not always well informed about what it is that they are purchasing or why they are purchasing one item rather than another. Listing the relative benefits of each option is thus part of the marketing function. It makes the client better informed about what is on offer.

Just as not all benefits are tangible, neither are they always easy to express in monetary values. It is difficult, for example, to assess any benefits of using one particular finish rather than another, in terms of increased company prestige. In the absence of any such measurements, professional value judgement can be used to rank each option with respect to relative benefits.

Quantifying the revenue potential of a particular choice

is another area where professional judgement is valuable. Suppose that the project is a private-sector one, in which a property developer is considering alternative wall and floor finishes for flats. The developer wants to know the relative increase in rent or sale value associated with each option. Will one finish generate a higher rent than another? Professional judgement will give the answer, or at least a ranking of the options available.

Once all the relevant and practically-obtainable benefits have been measured in their own units or presented in terms of professional value judgement, they need to be brought together so that they can be compared with one another, and with the costs. At this stage a broad distinction must be made between those benefits which can be quantified in monetary terms and those which cannot. The former can be included in the life cycle costing exercise in a straightforward manner either as revenues or as negative costs, appropriately discounted. Other techniques are required to handle those benefits that cannot be quantified in monetary terms. One possible approach is to use these benefits to guide the selection of finishes to be included in the analysis. Just how this is done is discussed in the context of the weighted evaluation matrix presented in Chapter 6.

In certain situations it may not be possible to attempt any kind of quantification of some benefits; such intangible benefits should simply be listed in the same manner as intangible costs, and then should be included in subsequent sensitivity analyses.

Benefit measurement clearly depends on consultation. It may reflect trade-offs between, for example, prestige and profit; these trade-offs can be decided upon only by the client.

Benefits and costs combined

The next stage in the appraisal process is to bring the benefits and the costs together. The purpose of a life cycle costing exercise is to identify the option with the lowest net cost. Where relevant costs and benefits can be easily identified and quantified in monetary terms this is simple. The value of any benefits is subtracted from the total costs of each option and the least-cost finish is thus identified: but how should the analysis proceed when there are intangible costs and benefits which cannot be readily quantified in monetary terms?

The starting point of such an exercise is the list of unquantifiable benefits drawn up earlier in the appraisal. If these benefits are subjected to a rigorous type of benefit appraisal some quantification will be possible by the weighted evaluation already mentioned. This subjective 'points score' can be employed in the appraisal exercise in the following way. Suppose there are four internal finishes: A, B, C and D. The relevant information is presented in the decision matrix of Table 5.1.

The first point which emerges from this table is that finish D should not be chosen: although it is cheaper than both B and C it has a higher net cost than finish A but provides a lower level of those benefits which cannot be quantified in monetary terms. It would obviously be more sensible for the decision-maker to choose finish A, a cheaper option which offers a higher level of associated benefits. In the cases of finishes B and C the answer is less clear-cut. Both these finishes are more expensive than finish A, but they both have a higher level of associated benefits. The decision-maker is thus forced to consider whether or not the additional benefits are worth the extra cost. In the example given, the initial question would be whether or not the additional 10 units of benefit provided by finish C were worth the extra £20 that they would cost. If it was considered that they were, and finish C was chosen, the

Table 5.1 A decision matrix.

FINISH	QUANTIFIABLE NET COSTS (£)	BENEFIT VALUES Unquantifiable in Monetary Terms
A	100	105
B	150	125
C	120	115
D	110	95

second question would be whether or not the additional 10 units of benefit provided by finish B were worth the further £30 that they would cost. The final choice would thus rest between finishes B and C. If finish C was rejected at the previous stage of the appraisal, the final choice would arise out of a comparison between finishes A and B and the outcome would be determined by whether or not the decision-maker considered that the extra 20 units of benefit were worth the extra £50 that they would cost.

In summary, this kind of decision-matrix approach can provide unambiguous answers in those cases where one option costs either the same or more than another, but provides fewer benefits. The option with fewer associated benefits will always be rejected. The approach cannot, however, provide unambiguous answers when more expensive options are associated with a higher level of benefits. In this case the technique serves to structure the decision in such a way that the decision-maker is made aware of the trade-off between costs and benefits.

Before the final choice of finish is made, it is also necessary to examine the list of any unquantifiable costs drawn up earlier in the appraisal exercise. Relative benefits have been compared with relative quantifiable net costs, but no account has been taken of those costs which cannot be quantified in monetary terms. As a final check, these costs should be examined in order to see whether they might lead to a change in the decision already made. It was noted above that the question which the decision-maker would have to ask is 'how large would these unquantifiable costs have to be before they would change the decision?' If their magnitude would have to be much larger than might reasonably be anticipated, the final selection can be made with confidence. If the margin for error is relatively small, then much greater attention should be paid to attempts to estimate these costs.

One final point is that public-sector clients are not profit maximisers. They have limited budgets and so wish to maximise the benefits obtained from a fixed-cost outlay. Differences between public and private-sector clients will tend to show up in the scores that they assign to each benefit item: in the private sector, for example, the visual and aesthetic dimension might be more important than durability.

Chapter 6 *Preliminary selection of finishes*

In these days of the microcomputer, the life cycle cost of a particular finish is easily, cheaply and rapidly generated once all the data are available. This proviso is, however, of crucial importance. Life cycle costing is expensive in its data requirements – initial capital costs, maintenance, recurrent and replacement costs, maintenance cycles, expected lives, and replacement cycles must all be estimated. Even with an extensive database this can be a time-consuming exercise. It is essential, therefore, that such effort be expended efficiently. Life cycle costing is best seen as a two-stage exercise. In the first stage, the finishes which should be considered are identified: in the second, the life cycle costs of these finishes are calculated. This chapter concentrates on the first stage, which inevitably will be subjective, but, we shall see, can be founded on objective criteria.

This first stage chooses from all possible solutions to a particular 'finishes problem' those finishes that satisfy the technical, performance and aesthetic requirements. There is no point in preparing a full life cycle cost plan for a finish that more detailed initial investigation would have shown to be incapable of satisfying, for example, certain technical prerequisites inherent in the problem being addressed.

Technical, performance and aesthetic criteria will vary depending upon the use of the space to which the finish will be applied and upon the needs of the client. There is a bewildering array of potential finishes from which to choose. So how can these considerations be taken into account? Three classifications are needed:

- the function for which the space is to be used

- the criteria upon which the choice of finish is to be based

- the extent to which particular finishes satisfy these criteria.

Step 1: classification of functions

There are many wall and floor finishes suitable for use in a variety of situations: some examples are given in Figs 6.1 to 6.4. In order to classify and to choose between them, it is helpful first to set up a basic classification system for the use of the spaces to which the finishes are to be applied. This will immediately focus attention on the considerations that are most relevant to each particular function. Such a system can be based on specific technical criteria such as anti-corrosive requirements or ability to withstand heavy weights, or can be more broadly based on considerations of specific functions, for example as a kitchen, toilet or corridor.

TYPE OF FINISH	APPLICATION TO
1. Ceramic Tiles	● Blockwork ● Timber Partitioning ● Plaster ● Plasterboard
2. Ceramic Mosaic Tiles	● Blockwork ● Timber Partitioning ● Plaster ● Plasterboard
3. Wallpaper	● Plaster ● Plasterboard
4. Fabric Wallcovering	● Plaster ● Plasterboard ● Blockwork
5. Emulsion Paint	● Plaster ● Plasterboard ● Blockwork
6. Timber Panelling	● Blockwork ● Timber Framing
7. Plastic Profiled Sheeting	● Steel Framework

Fig. 6.1 Wall finishes – internal.

TYPE OF FINISH	APPLICATION TO:
1. Ceramic Tiles	● Cement Screed ● Concrete Subfloor
2. Ceramic Mosaic Tiles	● Cement Screed ● Concrete Subfloor
3. Stone (Natural)	● Concrete Subfloor
4. Timber Block	● Suspended Timber Subfloor ● Concrete Subfloor ● Cement Screed
5. Timber Boarding	● Suspended Timber Subfloor ● Concrete Subfloor ● Cement Screed
6. Carpet	● Suspended Timber Subfloor ● Concrete Subfloor ● Cement Screed
7. Carpet Tiles	● Suspended Timber Subfloor ● Concrete Subfloor ● Cement Screed
8. Thermoplastic Tiles	● Suspended Timber Subfloor ● Cement Screed
9. PVC Tiles	● Suspended Timber Subfloor ● Cement Screed
10. PVC Welded Sheet	● Suspended Timber Subfloor ● Cement Screed
11. Insitu Granolithic	● Concrete Subfloor
12. Insitu Terazzo	● Concrete Subfloor
13. Precast Terazzo Tiles	● Concrete Subfloor

Fig. 6.2   Floor finishes – internal.

14. Marble	● Concrete Subfloor
15. Quarry Tiles	● Concrete Subfloor ● Cement Screed
16. Cork Compound Sheet	● Suspended Timber Subfloor ● Cement Screed
17. Cork Compound Tiles	● Suspended Timber Subfloor ● Cement Screed

Fig. 6.2 *Continued.*

TYPE OF FINISH	APPLICATION TO:
1. Ceramic Tiles	● Blockwork ● Concrete ● Cement Rendering
2. Ceramic Mosaic Tiles	● Blockwork ● Concrete ● Cement Rendering
3. Cement Paint	● Rendering ● Concrete
4. Timber Panelling	● Blockwork
5. Facing Brickwork	● ————
6. Clay/Concrete Tiling	● Timber Framing ● Blockwork
7. Glass Reinforced Plastic	● Blockwork ● Concrete
8. Plastic Profiled Sheeting	● Steel Framework

Fig. 6.3 Wall finishes – external.

TYPE OF FINISH	APPLICATION TO:
1. Ceramic Tiles	● Concrete Sub-base ● Cement Screed
2. Ceramic Mosaic Tiles	● Concrete Sub-base ● Cement Screed
3. Stone (Natural)	● Concrete Sub-base
4. Insitu Granolithic	● Concrete Sub-base
5. Mastic Asphalt	● Concrete Sub-base
6. Quarry Tiles	● Concrete Sub-base
7. Insitu Terazzo	● Concrete Sub-base
8. Terazzo Tiles	● Concrete Sub-base
9. Marble	● Concrete Sub-base

Fig. 6.4   Floor finishes – external.

Designers and clients tend to think in functional rather than specifically technical terms. Since the designer must be involved closely in this first stage of the analysis, it follows naturally that a classification system based on function is the one that should be used. At the same time, this system should be related to the way in which data are collected and presented. This implies that the function-based system should be related to something like the CI/SfB system.

Such a combined system is intended to encourage the participants in the design and development process to think about the selection of wall and floor finishes in a broader, more systematic manner than now appears to be the case. Three steps link the design of various facilities through to the characteristics of each functional space within a building. The exercise is illustrated fully in the case studies presented in

Functional Space	Item	Characteristics
Meat Preparation Area	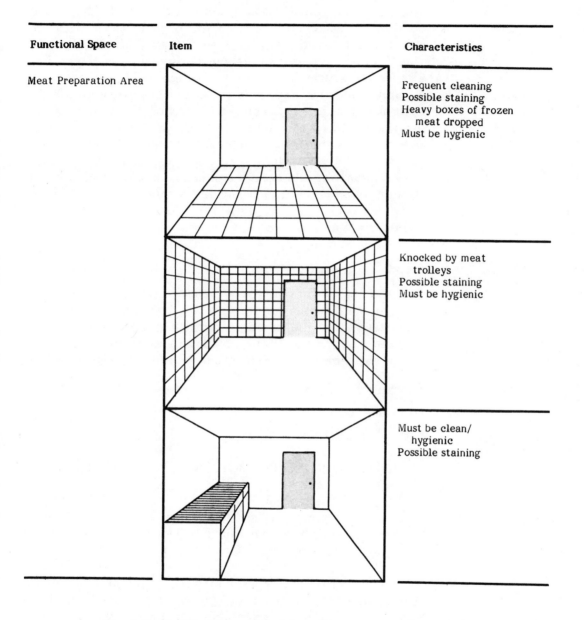	Frequent cleaning Possible staining Heavy boxes of frozen meat dropped Must be hygienic
		Knocked by meat trolleys Possible staining Must be hygienic
		Must be clean/ hygienic Possible staining

Fig. 6.5 Facility type: Administrative, commercial, protective services
Sub-category: Trading shops – supermarkets

Chapter 7, but may be briefly described here. It requires the completion of the kind of form presented in Fig. 6.5. The first step is the functional classification just described. The type of building is listed at the top of the page together with any relevant sub-classification, both from the CI/SfB classification. The form is completed by listing the functions of the various locations or spaces in column 1, the item being considered in column 2 and the associated characteristics in column 3. For example, if a supermarket is being considered, one of the functional spaces may be a meat preparation area. This has specific characteristics: it is likely to have heavy boxes of frozen meats dropped on the floor or knocked against the walls, to get dirty regularly and to need maintenance to a high standard of cleanliness and hygiene. Using the technique set out here a comprehensive listing of these basic functional requirements can be made in a systematic manner.

Step 2: choice criteria

The next step in the process is the identification of the criteria upon which the choice of finishes is to be made: the particular characteristics of the functional spaces found in the previous step must be translated into detailed requirements.

These criteria can be split into three broad types:

- technical performance
- visual/environmental
- initial/long term cost.

The detailed criteria under each of these broad headings are set out in Appendix A.

This part of the analysis works in a similar way to that for the first step: the format is shown in Fig. 6.6. Column 1 lists the requirements for each functional space as suggested by the characteristics in Step 1. Column 2 specifies the criteria the finishes must satisfy. Column 3 contains detailed specifications for the finishes such as types of material, colour and tolerances. At the end of this stage a list of precise requirements has been identified, which any acceptable finish must be capable of meeting.

Requirements	Criteria	Specification
Durability Water resistance Slip resistance Chemical resistance Acid resistance Alkali resistance Ease of cleaning Strength Stain resistance Hygiene	**Technical:-** Bacterial/ Microbial resistance Hardness Water absorption Moisture movement Compressive strength Transverse strength Slip resistance Chemical resistance **Visual:-** Stain resistance **Cost:-** Low cleaning Low maintenance	

Fig. 6.6   Facility type: Administrative, commercial, protective services
Sub-category: Trading shops – supermarkets
Functional space: Meat preparation area
Item: Floor

Step 3: selecting the finishes for appraisal

The final step identifies those finishes that should be included in the full life cycle costing exercise. The approach may be either simple or sophisticated: both have their merits, but this preliminary sift is essential. It consists of two stages. The first concentrates on technical criteria and the second on more subjective criteria, such as visual and aesthetic properties.

The initial consideration of solely technical/performance

criteria cannot be by-passed. There is no point whatsoever in including in the life cycle cost exercise a particular finish that does not meet the technical requirements specified. It does not matter if its visual and cost characteristics are highly satisfactory: if it fails to meet the technical specifications it simply cannot be used.

The initial selection can be made by using a simple checklist approach. The technical requirements for the finishes are checked against the technical specifications of the various finishes. Only those finishes that meet the technical requirements are considered further, as shown by Fig. 6.7. In this simple example only ceramic tiles and marble meet the technical specification, so only these should be considered further. When the finishes that are technically acceptable have been identified, the second stage brings the visual and cost criteria into the exercise. Again, this may be on a simple or sophisticated basis. In its most simple form, it would require a repetition of the checklist approach used above: the relevant visual and cost criteria would be listed and checked off against the characteristics of the various finishes, and those materials which satisfy all the relevant criteria would proceed to the next stage of the analysis.

Such a simple approach will not always be best. When a relatively 'undemanding' environment is being considered, it may well be that a large number of competing finishes are identified as being suitable for inclusion in the full life cycle costing, in which case the evaluation will become hopelessly complicated. In these circumstances, a more sophisticated approach is necessary to identify more precisely the relative merits of the various finishes as they relate to the particular functional space under consideration. When many finishes will satisfy some or all of the criteria at least to some extent, an approach that performs two basic functions is needed:

- it should identify which of the criteria that are declared as necessary are more important; for example, whether 'ease of cleaning' is more important than 'surface hardness';

- it should identify the extent to which each finish satisfies these criteria and, therefore, which finishes are acceptable for a full life cycle cost appraisal.

Criteria	Requirement	Ceramic Tiles	Vinyl Tiles	Granolithic	Marble
Abrasion resistance	▲	▲		▲	
Acid resistance	▲	▲	▲		
Alkali resistance					
Anti-static					
Bacterial/Microbial resistance	▲	▲			▲
Closed pores	▲	▲	▲		▲
Compressive strength	▲	▲			▲
Deep abrasion resistance	▲	▲		▲	▲
Electricity/Magnetism/ Radiation resistance					
Fire resistance					
Frost resistance					
Hardness					
Material alergies	▲	▲		▲	▲
Moisture movement	▲	▲		▲	▲
Slip resistance	▲	▲		▲	▲
Sound insulation					
Thermal conductivity					
Thermal expansion					
Thermal shock					
Transverse strength					
Water absorption	▲	▲	▲	▲	▲
Wear resistance	▲	▲		▲	▲
Weight					

Fig. 6.7 Selection of finishes – preliminary analysis.

Facility type: Administrative, commercial, protective services
Sub-category: Trading shops – supermarkets
Functional space: Meat preparation area
Item: Floor

Weighted evaluation techniques are the answer: Dell'Isola and Kirk (1981) have modified a management technique to develop the weighted evaluation technique. The various criteria are weighted numerically according to the importance that is attached to them and the finishes are scored according to the degree to which they meet the criteria. Finally, an overall score is calculated for each finish.

This weighted evaluation procedure works in the following way (see Fig. 6.8). Each of the criteria that the finish must satisfy are listed (A, B, C, etc.), in the criteria scoring matrix in the top half of the form. Each is then systematically compared with every other criterion in terms of direction and strength of preference. For example, for this particular functional space, 'water absorption' (criterion G) is given a minor preference over 'moisture movement' (criterion D) and so is assigned a score of 2 (G–2) in the appropriate cell of the criteria scoring matrix. Water absorption and slip resistance (criterion E) are ranked equally, giving the entry E/G, while bacterial/microbial resistance (criterion A) is given a major preference over every other criterion – the ranking A–4.

The process is repeated until all the criteria have been compared with one another and the results recorded in the appropriate cells of the matrix. The raw scores for each criterion are found by adding up the numbers in each cell where a preference for a particular criterion has been registered. Take, for example, criterion G. To obtain the raw score for G, all the cells in the two diagonals originating from G are examined. The upwardsloping diagonal contains G–3, E/G, G–2, G–3, G–3, A–4 and the downward sloping diagonal G–4, G–3. This gives a total score for G of 3 + 1 + 2 + 3 + 3 (+ 0) + 4 + 3 = 19. Note that the 'no preference' entry E/G scores 1 and A–4 scores zero.

This is done for every criterion to generate the raw score line in Fig. 6.8. This set of numbers identifies the relative importance of the various criteria in this particular functional space. Thus in this example bacterial/microbial resistance is very important, while stain resistance and low cleaning cost are of very minor importance relative to the other criteria.

These raw scores can now be used directly in the next stage of the exercise. However, for convenience, they may be converted into a set of weights on a scale of 1 to 10. To change

Criteria

Criteria Scoring Matrix

Importance

4 - Major preference
3 - Medium preference
2 - Minor preference
1 - Letter/Letter
No preference each scores one point

	Criteria										
A.	Bacterial/Microbial resistance	A-4									
B.	Compressive strength	B/C	A-4								
C.	Hardness	C/D	D-2	A-4							
D.	Moisture movement	E-2	E-2	E-2	A-4						
E.	Slip resistance	E-2	D/F	C-2	B/F	A-4					
F.	Transverse strength	G-3	E-2	G-2	G-3	G-3	A-4				
G.	Water absorption	G-4	E-2	E/G	D-3	B-2	B-2	A-4			
H.	Stain resistance	H-1	G-3	E-2	E-3	D-2					
I.	Low cleaning cost			F-2							

			H	G	F	E	D	C	B	A

Alternatives	Raw score	1	1	19	6	16	9	9	6	32	Total
Analysis Matrix	Weight (0–10)	0.5	0.5	6	2	5	3	3	2	10	
1. Ceramic Tiles		5 / 2.5	5 / 2.5	5 / 30	4 / 8	4 / 20	4 / 12	4 / 12	4 / 8	5 / 50	145
2. Vinyl Tiles		4 / 2	4 / 2	4 / 24	3 / 6	2 / 10	3 / 9	1 / 3	1 / 2	2 / 20	78
3. Granolithic		3 / 1.5	3 / 1.5	3 / 18	5 / 10	3 / 15	3 / 9	3 / 9	2 / 4	3 / 30	98
4. Marble		3 / 1.5	5 / 2.5	5 / 30	5 / 10	2 / 10	5 / 15	5 / 15	5 / 10	5 / 50	144

Excellent 5; Very Good 4; Good 3; Fair 2; Poor 1.

Fig. 6.8 Weighted evaluation.

Facility type: Administrative, commercial, protective services
Sub-category: Trading shops – supermarkets
Functional space: Meat preparation area
Item: Floor

raw scores R, into weights W, use the formula:

$$W = \frac{10R}{M(n{-}1)} \tag{30}$$

where M is the maximum score for each criterion (4, in the example) and n is the number of criteria, 9 in the present case. Thus in Fig. 6.8: $W = 10R/4(9{-}1) = 0.3125R$,

A raw score of 16 for criterion E then gives a weight of $16 \times 0.3125 = 5$, as entered in the weight line of Fig. 6.8. The advantage of this is that the second stage of the work may be undertaken without a calculator, no matter how large the criteria matrix.

Whichever approach is adopted, this second stage requires each finish to be scored (on a range of 1 to 5) according to the degree to which it satisfies each of the criteria set out in the matrix. These scores are inserted in the top box of each cell of the analysis matrix, beneath the criteria matrix. Thus, for example, ceramic tiles are rated as 'excellent' (score of 5) on criteria A, G, H and I, and 'very good' (score of 4) on the remainder. Vinyl tiles are rated highly on criteria G, H and I, but as 'poor' or 'fair' on criteria A, B, C, D and E.

These scores are then multiplied by the weights (or, if preferred, the raw scores) to give the entries in the lower box of each cell of the analysis matrix. For example, ceramic tiles are given a score of 30 (5×6) for criterion G (water absorption) and 50 (5×10) for criterion A (bacterial/microbial resistance).

The overall score for each finish is computed simply by adding the scores in the lower boxes across the whole row of criteria. It is then possible to proceed to the life cycle costing exercise by selecting, for example, only the three top scoring finishes, or only those that score above a predetermined level.

In summary, what the weighted evaluation of Fig. 6.8 does is firstly to identify the relative importance of the various criteria which the choice of finish has to satisfy in the particular functional space. Secondly, it shows the extent to which the various finishes satisfy these criteria, both individually and in aggregate. Most finishes are rated highly on at least one criterion, for instance, the granolithic finish has excellent transverse strength, but this is relevant only in a particular functional space.

Chapter 7 Using the life cycle approach and techniques in practice

This chapter presents two hypothetical case studies to illustrate how the approach and techniques already explained work in practice. It should be stressed that the contents of the following matrix tables are illustrative, not exhaustive. These examples are hypothetical but the life cycle cost comparisons are based upon actual cost data.

Supermarket sales area floor

The first study involves the selection of a floor finish for the general sales area of a supermarket. Figure 7.1 demonstrates how the hierarchical approach described in Chapter 6 may be used to identify the particular characteristics associated with the floors within major functional spaces of a supermarket sales area. The intention here is to obtain a list of characteristics from which an inventory of requirements appropriate to both wall and floor finishes may be drawn up. This is illustrated in Fig. 7.2.

Consider a particular functional area such as the entrance area. The characteristics listed in the final column of Fig. 7.1 for the floor and walls of the entrance area are translated into the basic requirements listed in the first column of Figs 7.2 and 7.3. These are translated into the detailed criteria presented in column 2. The final column in each table is reserved for recording any precise specifications.

The process is carried a stage further for the floor finish in Fig. 7.4, in which the required technical criteria are listed against a range of floor finishes to see which floor finishes meet the requirements. On the basis of this analysis, ceramic tiles, terrazzo and marble pass through to be considered in the full life cycle costing exercise given in Fig. 7.5.

The results of the exercise are straightforward. The initial capital costs of each option are set down together with an

Functional Space	Item	Characteristics
Entrance Area	Floor	• Heavy traffic • Wet in bad weather • Dirt brought in from outside • Must be flat • Design must fit company identity
	Walls	• Knocked by trolleys • Possible design feature • Likely to get dirty • Design must fit company identity
General Selling Area	Floor	• Relatively heavy traffic • Some spillage likely • Must be flat for trolleys • Heavy objects may be dropped on it • Company colour required • Easy maintenance/replacement
	Walls	• No design statement required • Likely to get relatively dirty • May be knocked by trolleys in places
	Columns/ Pillars	• No design statement required • Knocked by trolleys • Likely to get dirty
	Counter ends	• Should be unobstructive • May be knocked by trolleys
Delicatessen	Floor	• Must be hygienic • Wash frequently • Possible design feature • Some spillage likely
	Walls	• High standard of cleanliness • Possible design statement

Fig. 7.1   Facility type: Administrative, commercial, protective services
Sub-category: Trading shops – supermarkets

Functional Space	Item	Characteristics
Delicatessen (cont.)	Work Surfaces	• High standard of cleanliness • Likely to get wet • May be liable to staining
Wet Fish Area	Floor	• Generally wet environment • Must be hygienic/clean • Possible design feature
	Walls	• As for floor
	Work surfaces	• Hygienic • General cleanliness • Wet
Fresh Meat Area	Floor	• Must be hygienic • Frequent cleaning • Possible staining • Possible design feature
	Walls	• As for floor
	Work surfaces	• Hygiene/cleanliness • Frequently washed
Patisserie	Floor	• Cleanliness • Design statement
	Walls	• As for floor
	Surfaces	• Scratched by bread • Must be hygienic • Easily cleaned
Check-Out Area	Floor	• Heavy traffic • May get wet/dirty from entrance • Must be flat
	Walls	• Frequently knocked • Attractive design required • Likely to get dirty

Fig. 7.1   *Continued.*

Functional Space	Item	Characteristics
Meat Preparation Area	Floor	• Frequent cleaning • Possible staining • Heavy boxes of frozen meat dropped • Must be hygienic
	Walls	• Knocked by meat trolleys • Frequent cleaning • Possible staining • Must be hygienic
	Work surfaces	• Must be hygienic/clean • Possible staining
Fruit Preparation Area	Floor	• Frequent cleaning • Possible staining • Hygienic • Boxes may be dropped
	Walls	• Clean/hygienic • Likely to get knocked
	Work surfaces	• Hygienic/clean • Possible staining
Cold Room	Floor	• Must be hygienic/clean • Possible staining • Frost resistance required • May get knocked
	Walls	• As for floor
Medical Room	Floor	• Hygienic/clean
	Walls	• Hygienic/clean
	Wash basin splashback	• Likely to get wet • Likely to get dirty/stained

Fig. 7.1   *Continued.*

Functional Space	Item	Characteristics
Kitchen	Floor	• Must be hygienic/clean • Spillages likely • Possible staining
	Walls	• Clean/hygienic • Likely to be knocked
	Work surfaces	• Hygienic/clean • Possible staining
Lavatories	Floor	• Frequent cleaning • Wet environment
	Walls	• As for floor
	Wash basin splashback	• Likely to get wet • Likely to get dirty/stained
Storage Areas	Floor	• Likely to get knocked • Frequent cleaning • Strong enough for fork lift truck • Some spillage • Dirt from outside during deliveries
	Walls	• Likely to get knocked • Some dirt
Offices	Floor	• Relatively heavy traffic • Sound proofing • Some degree of comfort • Good visual aspects for public offices • Thermal insulation
	Walls	• Sometimes knocked • Contribution to pleasant working environment • Good design - especially if seen by public • Insulation - sound and heat

Fig. 7.1 *Continued.*

Requirements	Criteria	Specification
Durability Water resistance Slip resistance Ease of cleaning Flatness Ease of repairs/ maintenance Ease of replacement Appropriate design	**Technical:-** Hardness Water absorption Abrasion resistance Deep abrasion resistance Slip resistance Wear resistance Moisture movement **Visual:-** Size Shape Colour Flatness Variation **Cost:-** Low repair/ maintenance Low cleaning Low replacement	

Fig. 7.2   Facility type: Administrative, commercial, protective services
Sub-category: Trading shops – supermarkets
Functional space: Entrance area
Item: Floor

Requirements	Criteria	Specification
Durability Water resistance Ease of cleaning Ease of repairs/ maintenance Ease of replacement Appropriate design	**Technical:-** Hardness Abrasion resistance Deep abrasion resistance Compressive strength **Visual:-** Size Shape Colour **Cost:-** Low repair/ maintenance Low cleaning Low replacement	

Fig. 7.3 Facility type: Administrative, commercial, protective services
Sub-category: Trading shops – supermarkets
Functional space: Entrance area
Item: Floor

	Abrasion Resistance	Acid Resistance	Alkali Resistance	Anti-Static	Bacterial/Microbial Resistance	Closed Pores	Compressive Strength	Deep Abrasion Resistance	Electricity/Magnetism/ Radiation Resistance	Fire Resistance
Linoleum	◁									
Cork Tiles										
Quarry Tiles	◁									
PVC Welded Sheet										
Marble	◁									
Terrazzo	◁									
Granolithic										
Vinyl Tiles	◁									
Thermoplastic Tiles										
Carpet										
Wood Strip and Board										
Wood Block										
Ceramic Tiles	◁									
Required	◁									
Criteria										

Fig. 7.4 Facility type: Administrative, commercial, protective services
Sub-category: Trading shops – supermarkets
Functional space: Entrance area
Item: Floor

Criteria	Frost Resistance	Hardness	Material Allergies	Moisture Movement	Slip Resistance	Sound Insulation	Sub-Bedding/Bedding Requirements	Thermal Conductivity	Thermal Expansion	Thermal Shock	Transverse Strength	Water Absorption	Wear Resistance	Weight
Linoleum	◁			◁	◁							◁	◁	
Cork Tiles	◁				◁								◁	
Quarry Tiles	◁				◁								◁	
PVC Welded Sheet	◁			◁	◁							◁		
Marble	◁			◁	◁							◁	◁	
Terrazzo	◁			◁	◁							◁	◁	
Granolithic	◁												◁	
Vinyl Tiles	◁			◁	◁							◁	◁	
Thermoplastic Tiles	◁			◁								◁	◁	
Carpet	◁				◁									
Wood Strip and Board														
Wood Block														
Ceramic Tiles	◁			◁	◁							◁	◁	
Required	◁			◁	◁							◁	◁	

Fig. 7.4 *Continued.*

PROJECT LIFE: 10 years
Discount rate: 4%

Costs	Option 1 Ceramic Tiles Finish Life 10 years Est. costs	Pres. value	Option 2 Terrazzo Finish Life 10 years Est. costs	Pres. value	Option 3 Marble Finish Life 10 years Est. costs	Pres. value	Option 4 Finish Life Est. costs	Pres. value
Capital								
Floor screed (9.49/m2)		9490		9490		9490		
Ceramic Tiles (19.22/m2)		19220						
Terrazzo(24.51/m2)				24510				
Marble (69.74/m2)						69470		
Contingencies @ 5%		1436		1700		3948		
Total capital costs		30146		35700		82908		
Annual maintenance costs								
Ceramic Tiles Cleaning − 8.111	600	4867						
Terrazzo Cleaning − 8.111			750	6083				
Marble Cleaning − 8.111					650	5272		
Total annual maintenance costs		4867		6083		5272		
Year PV factor								
Maintenance/replacement/ alterations (intermittent)								
Total Maintenance/Replacement/ Alterations Costs		———		———		———		
Total Running Costs		4867		6083		5272		
Tangible Benefits								
Total Additional Tax Allowances		———		———		———		
Salvage and Residuals								
Total Salvage and Residuals		———		———		———		
Total Net Present Value of Life Cycle Costs		35013		41783		88180		

Fig. 7.5   Facility type: Administrative, commercial, protective services
Sub-category: Trading shops – supermarkets
Functional space: Entrance area
Item: Floor
Area: 1000 m^2

allowance for contingencies. The various items of initial capital cost are summed and details of all recurrent future costs listed. In this particular example, running cost items such as energy consumption together with other items of life cycle cost such as residual values are not applicable. Replacement costs are not considered, as the lives of the materials are well in excess of the period of analysis, taken as the anticipated replacement cycle of the functional space.

In this example, the use of life cycle costing techniques does not result in the selection of a different finish from that which would have been chosen on the basis of initial capital costs alone. The more expensive finishes included in this example also have higher associated future recurrent costs. These serve to reinforce the initial capital cost differences.

A life cycle cost profile for each floor finish is shown in Fig. 7.6. This is helpful in viewing at a glance the cost profiles of each floor finish over the period of analysis for the entrance area. Life cycle cost profiles require a large amount of iterative calculation of total NPV for each point in time and a computer-based model has been used for this purpose.

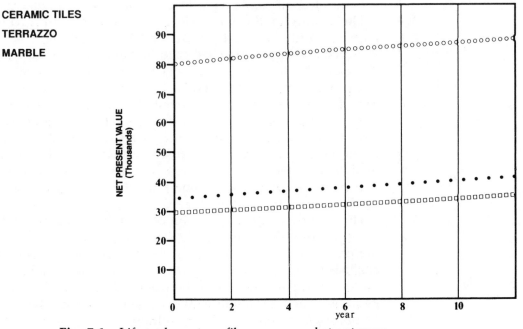

□ CERAMIC TILES
● TERRAZZO
○ MARBLE

Fig. 7.6 Life cycle cost profile – supermarket entrance.

School kitchen walls

This example involves the selection of a wall finish for a school kitchen. In this case it is assumed that only the school kitchens are being refurbished.

The first step of the exercise is presented in Fig. 7.7 and the relevant characteristics of school kitchen walls are briefly presented in the final column of this table. These are translated into the detailed requirements listed in the first column of Fig. 7.8 and the criteria that any acceptable finish must satisfy are listed in column 2. The last column is reserved in the normal manner for any detailed specifications. In Fig. 7.9 a selection of possible wall finishes is assessed against the technical criteria from Fig. 7.8 to determine whether or not their specifications satisfy the listed criteria. From Fig. 7.9 it can be seen that seven finishes have been identified as suitable for inclusion in the life cycle costing exercise. To enable the calculation stage of the exercise to be manageable a weighted evaluation of the seven acceptable finishes is carried out in Fig. 7.10. On the basis of this weighted evaluation three wall finishes are selected for inclusion in the calculation stage: ceramic tiles, oil paint on plaster and emulsion paint on plaster.

The results of the life cycle costing exercise are given in Figs 7.11 and 7.12. Both initial capital and running costs are included and appropriately discounted to obtain a total NPV at 30 years. Again, a life cycle cost profile has been generated using a computer-based model to show the performance of each finish over the time horizon. It can be seen from this graph that although in this case ceramic tiles are initially the most expensive wall finish, after 20 years ceramic tiles become, in life cycle terms, the most cost-effective. The difference in total NPV between each finish at 30 years can clearly be seen, and crossover points when one finish becomes more, or less, cost-effective can be identified.

This is a particular application of sensitivity analysis, which identifies the impact of variation in the period of analysis on the choice of finish. It emerges clearly that for any period of analysis in excess of about 18 years, ceramic tiles would be the preferred finish. The choice between oil paint and emulsion paint is much more difficult to make. Indeed, this figure emphasises the danger of concentrating on a single value for the period of analysis without any sensitivity analysis: if five

Functional Space	Item	Characteristics

Kitchen

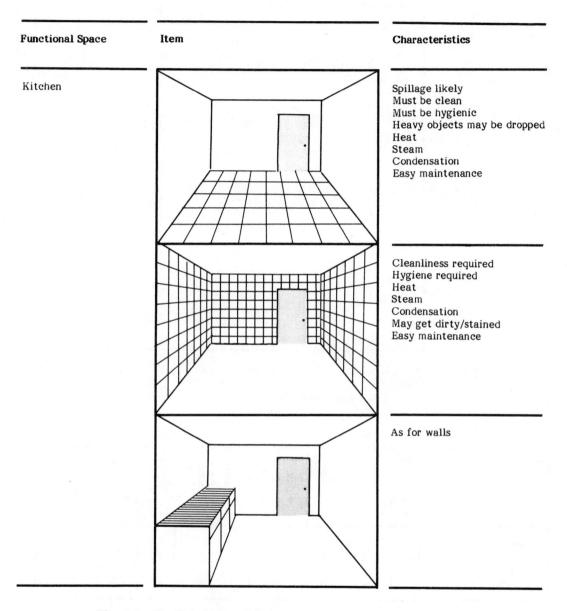

Spillage likely
Must be clean
Must be hygienic
Heavy objects may be dropped
Heat
Steam
Condensation
Easy maintenance

Cleanliness required
Hygiene required
Heat
Steam
Condensation
May get dirty/stained
Easy maintenance

As for walls

Fig. 7.7 Facility type: Education, scientific, information
Sub-category: Schools
Functional space: Kitchen

Requirements	Criteria	Specification
Hygiene Ease of cleaning Water resistance Stain resistance Ease of maintenance	**Technical:-** Bacterial/ Microbial resistance Water absorption **Visual:-** Stain resistance **Cost:-** Low cleaning Low maintenance	

Fig. 7.8   Facility type: Education, scientific, information
Sub-category: Schools
Functional space: Kitchen
Item: Walls

Criteria	Ceramic Tiles	Emulsion Paint	Cement Paint	Wallpaper	Fabric	Plywood Panelling	Facing Brickwork	Sprayed Texture	Plastic Laminate	Oil Paint on Render	Oil Paint on Plaster
Bacterial/Microbial Resistance	⚠	⚠	⚠					⚠	⚠	⚠	⚠
Water Absorption	⚠	⚠	⚠					⚠	⚠	⚠	⚠

Fig. 7.9   Facility type: Education, scientific, information
Sub-category: Schools
Functional space: Kitchen
Item: Walls

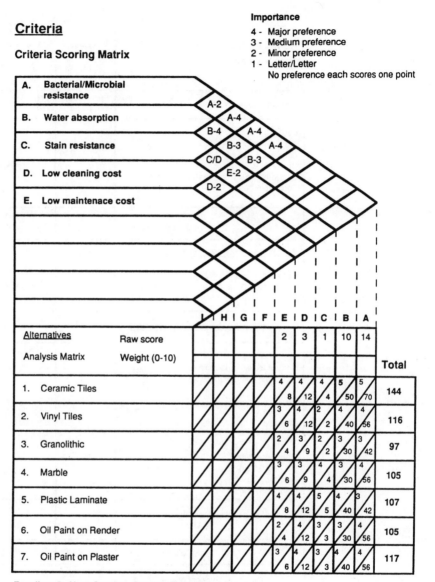

Fig. 7.10 Weighted evaluation.

Facility type: Education, scientific, information
Sub-category: Schools
Functional space: Kitchen
Item: Walls

Costs	Option 1 Ceramic Tiles Finish Life 30 years		Option 2 Oil Paint on Plaster Finish Life 7 years		Option 3 Emulsion Paint on Plaster Finish Life 5 years		Option 4 Finish Life	
	Est. costs	Pres. value	Est. costs	Pres. value	Est. costs	Pres. value	Est. costs	Pres. value
Capital								
Ceramic Tiles (£12/m2)	12000							
Oil Paint on Plaster (£4/m2)			4000					
Emulsion Paint on Plaster(£2.50/m2)					2500			
Contingencies @ 5%	600		200		125			
Total capital costs	12600		4200		2625			
Annual maintenance costs								
Ceramic Tiles (£0.30/m2/pa) - 17.292	300	5188						
Oil Paint (£0.50/m2/pa) - 17.272			500	8646				
Emulsion Paint(£0.60/m2/pa) -17.292					600	10375		
Total annual maintenance costs		5188		8646		10375		

Maintenance/replacement/ alterations (intermittent)	Year	PV factor							
Oil Paint	7	0.7599		4000	3040				
Oil Paint	14	0.5775		4000	2310				
Oil Paint	21	0.4388		4000	1755				
Oil Paint	28	0.3335		4000	1334				
Emulsion Paint	5	0.7835				2500	1959		
Emulsion Paint	10	0.6756				2500	1689		
Emulsion Paint	15	0.5553				2500	1388		
Emulsion Paint	20	0.4564				2500	1141		
Emulsion Paint	25	0.3751				2500	938		
Total Maintenance/Replacement/ Alterations Costs					8439		7115		
Total Running Costs		5188		17085		17490		
Total Net Present Value of Life Cycle Costs		17788		21285		20115		

PROJECT LIFE: 30 years
Discount rate: 4%

Fig. 7.11   Facility type: Education, scientific, information
Sub-category: Schools
Functional space: Kitchen
Item: Walls
Area: 1000 m^2

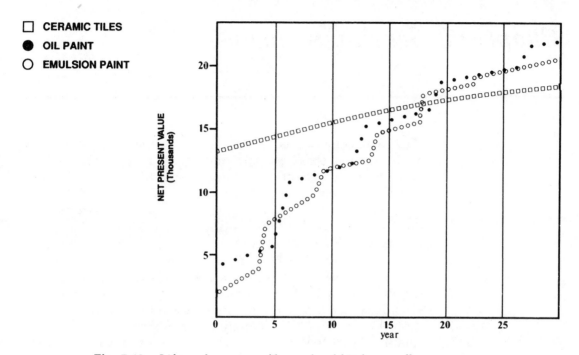

Fig. 7.12   Life cycle cost profile – school kitchen walls.

years had been used as the period of analysis a totally different answer would have been obtained from that generated by a seven-year period of analysis. The reason for that high degree of sensitivity also emerges clearly from Fig. 7.12. The different replacement cycles of oil and emulsion paints have a sharp impact on their NPVs and thus on their relative rankings over time.

Chapter 8 *Sensitivity analysis*

Risk management has already been discussed and sensitivity analysis considered as part of a general risk management system. In this chapter, sensitivity analysis is considered in rather more detail in the specific context of the life cycle costing examples of Chapter 7. An alternative method of presenting the sensitivity analysis results discussed in that chapter will be further developed here, for two reasons. Firstly, there is no uniquely 'correct' method of performing a sensitivity analysis. Chapter 4 presents one technique, but it is valuable to contrast it with this alternative technique. Secondly, where possible, sensitivity analysis should take advantage of the specific problem structure. It is quite clear from the examples in Chapter 7 that the period of analysis is a critical variable in the evaluation of different finishes. The sensitivity analysis should take account of this feature. There is the further benefit that bivariate sensitivity analysis is obtained rather than the standard single variable analysis.

Reasons for performing sensitivity analysis were extensively described in Chapter 4, but it is worthwhile briefly reviewing the discussion here. Any life cycle costing exercise is affected by the estimates and assumptions made. The future is being forecast on the basis of current data and knowledge. In addition, the degree to which a particular life cycle cost is affected by changes in these estimates and assumptions is likely to vary quite markedly, both across the options being compared, and with respect to the specific estimates and assumptions.

No simple *a priori* judgement can be made of which estimates and assumptions will be particularly important: this depends very much on the specific project. However, it is possible to identify more extreme cases; for example, where there is a wide disparity in both running costs and initial capital costs across the options being compared, both the period of analysis and discount rate will become particularly important; where the replacement cycle varies markedly, the estimate of that cycle becomes important, as does the period of analysis. Inflation can be expected to act differentially across options that exhibit markedly different cash flow profiles.

The various finishes satisfy many of these extreme conditions. This chapter concentrates, therefore, on sensitivity analysis of the life cycle costing calculations:

- the period of analysis
- the discount rate
- the life expectancy of the options
- the various cost estimates
- the rate of inflation.

The period of analysis

In the discussion of the general life cycle costing method it was noted that an assumption must be made about the period of analysis (p.35) of the project; a period which may not be known with any accuracy. Uncertainty stems from many different sources; in particular, buildings are affected by many kinds of obsolescence.

With the complete life cycle costing model it is a simple matter to construct profiles showing how the present value of the costs of various finishes varies over time. This was demonstrated in Fig. 7.12, which is repeated here as Fig. 8.1.

For each finishes option the net present value is calculated, assuming a discount rate of 4%, while allowing the period of analysis to vary continuously from 1–30 years. Thus, if the period of analysis is 10 years, the NPVs are approximately:

- emulsion paint   £12/m^2
- oil paint   £12/m^2
- ceramic tiles   £15.50/m^2.

If the period of analysis is 20 years the NPVs are:

- emulsion paint   £18/m^2
- oil paint   £18.50/m^2
- ceramic tiles   £17/m^2.

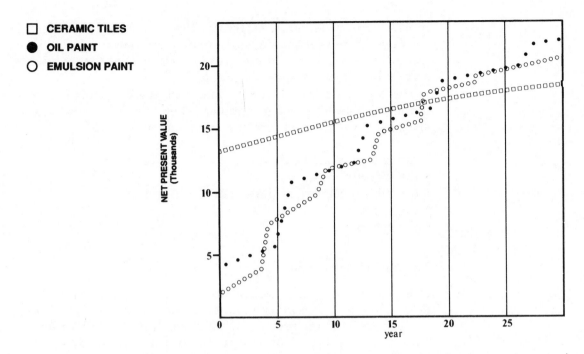

Fig. 8.1　Life cycle cost profile – school kitchen walls.

If non-quantifiable benefits are ignored, ceramic tiles are, in this example, the low-cost option for any period of analysis in excess of 20 years. With much shorter time horizons, emulsion paint or oil paint should be preferred on purely cost terms. A preference for ceramic tiles would have to be based on aesthetic and other more intangible grounds. Figure 8.1 emphasises the central importance of the period of analysis. The remaining sensitivity analyses take account of this feature.

The discount rate

The effect of varying the discount rate can be illustrated by examining the effect of increasing the discount rate from 4% to 10% in the second case study of Chapter 7: the relevant figures are presented in Fig. 8.2 and illustrated graphically in Fig. 8.3. An increase in the discount rate of this magnitude results in the emulsion paint option becoming the most economic finish for any period of analysis of more than about 6 years. Athough 10% is hardly realistic as an inflation-free (real) discount rate, it

		Option 1 Ceramic Tiles Finish Life 30 years		Option 2 Oil Paint on Plaster Finish Life 7 years		Option 3 Emulsion Paint on Plaster Finish Life 5 years		Option 4 Finish Life	
PROJECT LIFE: 30 years **Discount rate: 10%**									
Costs		Est. costs	Pres. value	Est. costs	Pres. value	Est. costs	Pres. value	Est. costs	Pres. value
<u>Capital</u>									
Ceramic Tiles (£12/m2)		12000							
Oil Paint on Plaster (£4/m2)				4000					
Emulsion Paint on Plaster(£2.50/m2)						2500			
Contingencies @ 5%			600		200		125		
Total capital costs			12600		4200		2625		
<u>Annual maintenance costs</u>									
Ceramic Tiles (£0.30/m2/pa) – 9.4269		300	2828						
Oil Paint (£0.50/m2/pa) – 9.4269				500	4713				
Emulsion Paint(£0.60/m2/pa) –9.4269						600	5656		
Total annual maintenance costs			2828		4713		5656		
Maintenance/replacement/ alterations (intermittent)	Year	PV factor							
Oil Paint	7	0.5132		4000	2053				
Oil Paint	14	0.2633		4000	1053				
Oil Paint	21	0.1351		4000	540				
Oil Paint	28	0.0693		4000	277				
Emulsion Paint	5	0.6209				2500	1552		
Emulsion Paint	10	0.3855				2500	964		
Emulsion Paint	15	0.2394				2500	599		
Emulsion Paint	20	0.1486				2500	231		
Total Net Present Value of Life Cycle Costs					3923		3718		
Total Running Costs			2828		8636		9374		
Total Maintenance/Replacement/ Alterations Costs			15428		12836		11999		

Fig. 8.2 Facility type: Education, scientific, information
Sub-category: Schools
Functional space: Kitchen
Item: Walls
Area: 1000 m^2

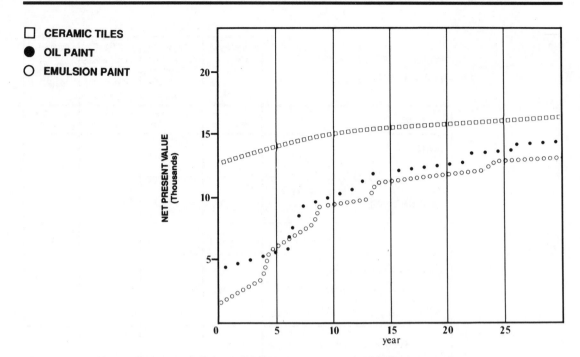

CERAMIC TILES
OIL PAINT
EMULSION PAINT

Fig. 8.3   Life cycle cost profile – school kitchen walls.
10% discount rate

illustrates a general point: the higher the discount rate chosen, the less the weight attached to the future costs included in the calculation and the greater the weight attached to initial capital costs. Lower discount rates have the reverse effect by increasing the relative importance of future costs. Put another way, a 'low capital cost' methodology is equivalent to choosing a comparatively high discount rate.

This point emerges particularly clearly by comparison of the cost profiles of figures 8.1 and 8.3. The higher discount rate offsets totally the longer-term advantages of ceramic tiles, and emphasises the short-term advantages of emulsion paint. There is now no time horizon in this example over which ceramic tiles would be preferred on cost terms, while the marginal advantage of emulsion paint over oil paint in Fig. 8.1 is further emphasised by the higher discount rate.

What would happen if the discount rate were lower? Reducing the discount rate to 2% generates the results in Fig. 8.4, giving the higher running costs of the paint finishes an even greater significance. At any project life in excess of

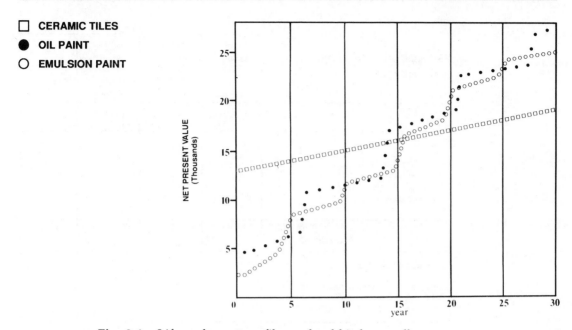

Fig. 8.4 Life cycle cost profile – school kitchen walls.
2% discount rate

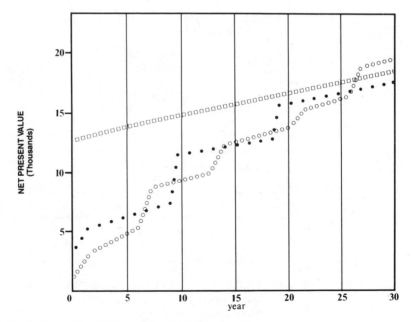

Fig. 8.5 Life cycle cost profile – school kitchen walls.
Increased finish life

	Option 1 **Ceramic Tiles** Finish Life 30 years		Option 2 **Oil Paint on Plaster** Finish Life 10 years		Option 3 **Emulsion Paint on Plaster** Finish Life 7 years		Option 4 Finish Life	
Costs	Est. costs	Pres. value	Est. costs	Pres. value	Est. costs	Pres. value	Est. costs	Pres. value
<u>Capital</u>								
Ceramic Tiles (£12/m2)		12000						
Oil Paint on Plaster (£4/m2)				4000				
Emulsion Paint on Plaster(£2.50/m2)						2500		
Contingencies @ 5%		600		200		125		
Total capital costs		12600		4200		2625		
<u>Annual maintenance costs</u>								
Ceramic Tiles (£0.30/m2/pa) – 17.292	300	5188						
Oil Paint (£0.50/m2/pa) – 17.272			500	8646				
Emulsion Paint(£0.60/m2/pa) –17.292					600	10375		
Total annual maintenance costs		5188		8646		10375		
Maintenance/replacement/ alterations (intermittent)	Year	PV factor						
Oil Paint	10	0.6756	4000	2702				
Oil Paint	20	0.4564	4000	1826				
Emulsion Paint	7	0.7599			2500	1900		
Emulsion Paint	14	0.5775			2500	1444		
Emulsion Paint	21	0.4388			2500	1097		
Emulsion Paint	28	0.3335			2500	834		
Total Maintenance/Replacement/ Alterations Costs				4528		5275		
Total Running Costs		2828		9241		10931		
Total Net Present Value of Life Cycle Costs		17788		17374		18275		

PROJECT LIFE: 30 years
Discount rate: 4%

Fig. 8.6 Facility type: Education, scientific, information
Sub-category: Schools
Functional space: Kitchen
Item: Walls
Area: 1000 m^2

15 years, the ceramic tile finish is more economic. In other words, the lower the discount rate, the more choice will be biased in favour of longer-term objectives. A low discount rate is equivalent to increasing the present value of future cost commitment.

Life expectancy of the options

The choice of finish can be affected by variations in the life of the finishing materials themselves. In the school kitchen walls case study, what would be the effect if the expected service life of the paint finishes were actually longer in practice than was assumed in the original calculations? An exercise assuming that oil paint on plaster will last 10 years and emulsion paint on plaster 7 years is shown in Figs 8.5 and 8.6. As might be expected, the changed assumptions improve the relative cost advantage of the paint finishes: a pure cost comparison favours one or other of the paint options at any period of analysis less than 25 years.

This can be contrasted with the results of Fig. 8.7, in which

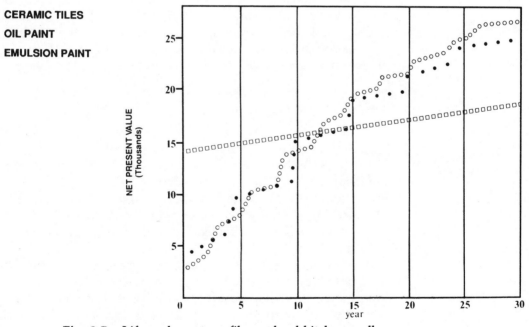

☐ CERAMIC TILES
● OIL PAINT
○ EMULSION PAINT

Fig. 8.7 Life cycle cost profile – school kitchen walls.
Reduced finish life

shorter lives are assumed for the two paint finishes. These finishes should now be chosen only if the period of analysis is less than 15 years. If the life expectancy of ceramic tiles was also reduced, say to 20 years, there would be a jump in the ceramic tile NPV curve at 20 years. Any period of analysis between 15 and 20 years would still result in ceramic tiles being the low-cost option. A 'grey area' would emerge between 20 and 25 years over which the three finishes would be roughly comparable in cost terms, but at anything in excess of 25 years ceramic tiles would re-emerge as the low-cost option.

It remains difficult to choose between the two paint finishes: the NPVs criss-cross with increased period of analysis as a consequence of the different repainting cycles, no matter what discount rate or replacement cycles are considered. This ambiguity was discussed in Chapter 7. It is valuable in emphasising the importance of sensitivity analysis and of the kinds of comparison being made here. The outcome of a single life cycle calculation is dependent upon the various assumptions underlying it. Only by illustrating the life cycle cost calculations, as in Fig. 8.5, can the source of the ambiguity be identified: in this case the impact of periodic replacement.

The final choice remains, of course, in the realm of managerial discretion, but the sensitivity analysis informs that choice in a particularly effective way.

The various cost estimates

The valuation of NPV of a particular option will be sensitive to a change in estimated costs: a 10% increase in all costs will increase NPV by 10%. In itself, however, this is of minor importance: it is the ranking of the various options that is important, rather than the absolute magnitude of their NPVs. This section, therefore, considers the extent to which ranking of finishes might be affected by a change in particular cost estimates.

If the cost increase is specific to a particular option, that option will become much less attractive; but to perform this kind of specific sensitivity analysis implies that there are some factors which are more likely to lead to uncertainty in the cost estimates of one finish rather than another. In the absence of such particular factors, the sensitivity analysis should

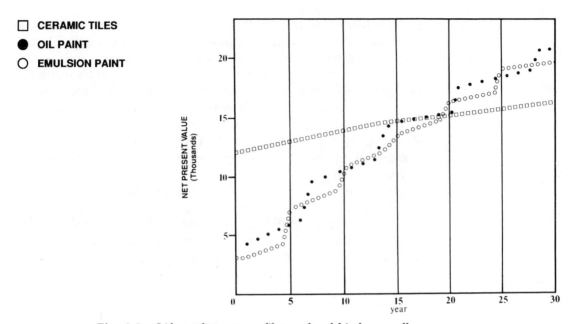

□ CERAMIC TILES
● OIL PAINT
○ EMULSION PAINT

Fig. 8.8 Life cycle cost profile – school kitchen walls.
Capital costs −10%

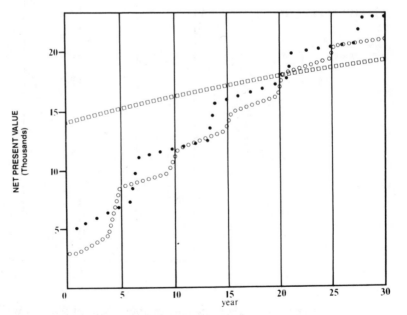

□ CERAMIC TILES
● OIL PAINT
○ EMULSION PAINT

Fig. 8.9 Life cycle cost profile – school kitchen walls.
Capital costs +10%

be conducted assuming that a specified class of costs has increased for *all* the options under consideration. It is this latter kind of sensitivity analysis that we examine here.

Comparison of Figs 8.1, 8.8 and 8.9 indicates that the ranking of the three finishes is affected very little by a 10% increase or decrease in capital cost estimates: 20 years remains the critical period of analysis. There is some indication that a general decrease in capital costs would favour the high capital-cost option, as might be expected, but this is little more than a marginal change.

The same result is obtained if running cost estimates are increased or decreased by 10%. The critical period of analysis remains at approximately 20 years although again, as should be expected, a general reduction in running cost estimates tends to favour the low-cost option: see Figs 8.10 and 8.11.

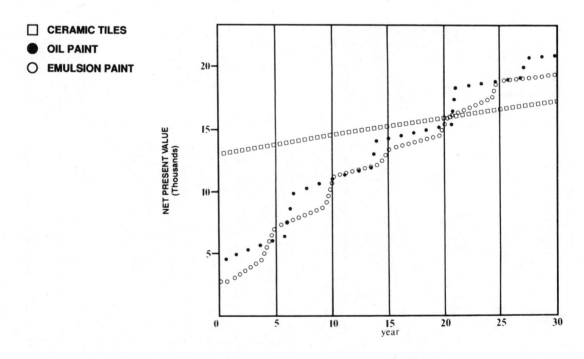

Fig. 8.10   Life cycle cost profile – school kitchen walls.
Running costs −10%

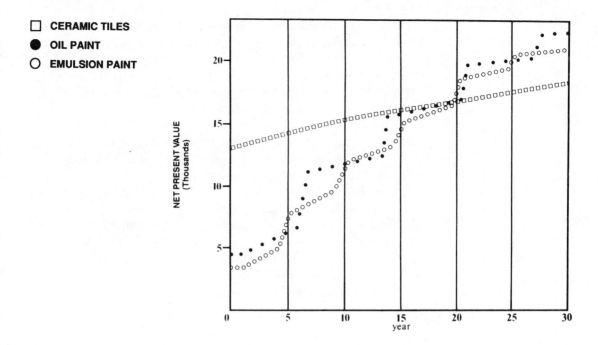

□ **CERAMIC TILES**
● **OIL PAINT**
○ **EMULSION PAINT**

Fig. 8.11   Life cycle cost profile – school kitchen walls.
Running costs +10%

The rate of inflation

Our examples so far have all assumed an inflation-free (real)
discount rate of 4% with any change being made for all options
under consideration. This is, in fact, the best way to handle the
general problem of inflation: Flanagan and Norman (1983)
discuss this in some detail. The central point is that all NPV
calculations must compare like with like. If costs are estimated
at today's prices, they must be discounted at an inflation-free
discount rate. If the cost estimates include a general price
adjustment for inflation, a nominal (or market) discount rate
including inflation should be used for discounting purposes.

Life is rather more complicated if different cost elements
inflate at different rates; for example, if labour costs (and so
maintenance and cleaning costs) are expected to increase at a
different rate from energy, initial capital, or material costs.
Where this is expected to be the case and when the differential
rates of inflation can be estimated, they should be explicitly

identified in the analysis. The simplest way to deal with such differential inflation rates is to continue to use an inflation-free discount rate to discount all future costs, but to add a differential escalation or deflationary factor to particular future costs, if these are expected to rise respectively by more or less than the general rate of inflation assumed in the analysis (see Chapter 2).

There remains, of course, the question of how such differential rates of inflation might be identified. The first and most obvious source is from published statistics. For example, the National Economic Development Office (NEDO) publishes indices for each of the initial capital cost elements of a building. Specific enquiries can be made for a more detailed analysis. Reference should also be made to the Retail Price Index (RPI), Construction Tender Price Index and Construction Cost Index to obtain a general view of inflation.

To convert an index to a percentage increase or decrease the following formula applies:

$$I = \frac{P_t - P_{t-1}}{P_{t-1}} \times 100 \tag{31}$$

Where:

I = rate of inflation for the unit period
P_t = latest index
P_{t-1} = previous index.

Care should be taken in computing rates of inflation from indices, as the period that each index represents may vary: for instance, monthly, quarterly or annual indices may be given. The calculated rate of inflation will then also vary and should be converted to an annual figure.

Taxation

All profitable companies are liable to taxation. Incorporating taxation into life cycle costing calculations can complicate the process. As a general rule expenditure is divided into two categories, capital and revenue. The distinction between capital and revenue must be understood.

Capital expenditure a term which is used to refer to money expended in acquiring assets, or in the permanent improvement of, or the addition to, or the extension of, existing assets which are intended for use in the carrying out of business operations. These assets should be expected to have a useful life of more than one year.

Revenue expenditure expenditure charged against expense in the period of acquisition. This term is used to refer to all expenditure which cannot be debited to an asset account. All the running costs of a facility tend to be revenue expenditure, which includes such items as rent, rates, fuel costs, cleaning and maintenance costs.

Revenue expenditure is offset in full against income before calculating taxable profit. Capital expenditure is treated differently and relief on capital costs is made through capital allowances which are set against the tax payable.

The Capital Allowances Act 1968 establishes the basis for computing capital allowances, though amendments have been made from time to time. The allowances are normally spread over a period of time. Under the 1988 Finance Act, plant and machinery is allowable on a 25% reducing balance, and industrial buildings on a 4% straight line basis.

The reducing balance allowance means that 25% of the capital cost can be claimed in the first year, in the second year it is 25% of the remaining 75% (i.e. 18.75% of the original cost), in the third year it is 25% of the remaining 56.25% of the original cost and so on.

The straight line allowance is 4% per annum of the original capital cost spread over 25 years. It must be emphasised that the writing down rates will not be the same as the way the company depreciates the asset in its balance sheet or for internal accounting. Depreciation, generally speaking, is not allowable against tax. Capital allowances effectively represent a type of accelerated depreciation.

Usually, payments are made a year in arrears, however the timing is complicated by Advanced Corporation Tax payable on dividends to shareholders. For most cash flows it is reasonable to assume that tax is paid on all revenue in the year following receipt of the revenue.

Chapter 9 *Case studies*

These case studies evaluate the total costs of a number of commonly used internal building finishes.

Classifying and analysing data

Eight different wall and floor finishes are taken to form a basic data set from which to construct the case studies. These are:

Floor finishes
- ceramic floor tiles
- natural stone
- hardwood timber strip
- carpet
- PVC tiles
- terrazzo tiles
- marble tiles
- quarry tiles.

Wall finishes
- ceramic wall tiles
- wallpaper
- fabric wallcovering
- emulsion paint
- hardwood timber wallboarding
- sprayed texture coating
- oil paint
- plastic faced wallboarding.

Initial capital cost data were collected from a number of live construction projects and a simple average taken of the adjusted data. A number of factors were included in this adjustment; for example, whether preliminary costs were included within each unit rate, and the location and type of project. These factors produced a set of data that reflects the mean or average price for each particular finish analysed. In adjusting the data set, the specification (both quantity and quality) of each finish was considered as most important in producing a set of comparable data for the life cycle calculations and analysis. The adjusted data set was therefore constructed to reflect an average quality and specification for each finish, allied with a quantity that would not adversely affect the price.

This is demonstrated by ceramic tiles, which range in quality and specification from a standard plain low-cost tile to a high quality and costly hand-made tile of individual design and size. The quantity fixed also has a major effect upon the final price: small quantities, such as may be expected for a minor works project, would normally significantly increase both material and labour costs, and large quantities would normally significantly decrease them.

The price data distributions for each group of finishes were further analysed to test whether individual price data seemed reliable. The initial cost data set was then checked against an elemental unit price rate build-up from first principles, using current labour and materials prices. All cost data were adjusted to a first-quarter 1986 base using a tender price index.

Initial installation costs for each finish include all necessary beds, backings and fixings, but where backings are common to all finishes, this cost has been omitted. For instance, for all floor finishings a lightweight cement and sand screed has been assumed as the backing material. Where differences do occur, these have been taken into account; for instance, it would be normal to lay a latex screed for PVC tiles and underlay for carpet.

Life cycle data on each of the finishes (such as maintenance, replacement and cleaning cycles) were collected from different sources, mainly from manufacturers and specialist contractors. The categories relating to different functional spaces were formulated, based upon the use of the space and also by the environmental criteria that each functional space demands of the surface finishes.

The breakdown of this classification system is based upon the environmental and physical conditions that exist in the functional space surrounding the surface. These conditions are translated into the following set of criteria:

- the physical wear sustained
- the cleanliness standard required
- the amount of water present.

Seven functional space categories have been identified for both floor and wall finishes. In the case studies these are environments A to G as shown in Table 9.1.

Table 9.1

Environment	Criteria		
	Physical wear	*Cleanliness*	*Water*
A	heavy	very clean	dry
B	heavy	average clean	dry
C	normal	very clean	dry
D	normal	average clean	dry
E	light	very clean	dry
F	light	average clean	dry
G	normal	highly clean	wet

For example, environment B for both floor and wall finishes comprises a heavy wear, average cleanliness and dry area, such as would be found in functional spaces such as an airport passenger terminal. Environment D, as a further example, comprises a normal wear, average cleanliness and dry area, such as would be found within a library or department store. Life cycle data were collected about the replacement, maintenance and cleaning of each finish, in each functional space category.

Replacement cycles represent the service life of the surface finish, while the associated replacement costs of the finishes were determined by making an appropriate adjustment to the initial capital cost to cover the costs of preparing the surfaces and working in small areas of an occupied building.

Annual, intermittent and regular maintenance data, were collected for each finish and also for each functional space category: maintenance requirements vary not only for each finish but also for the same finish in different functional spaces. Maintenance costs were then determined for each different finish, but standardised for all categories of functional space.

Cleaning cycles were determined and based upon daily, weekly and periodic cycles, depending upon the use and frequency of use of the building and the environmental considerations. The annual cleaning frequency was calculated for each functional space category but standardised for each finish. Cleaning costs, however, have been calculated for each different surface finish and standardised for each functional space category.

Calculating costs

Those finishes satisfying the criteria relating to each functional space category were selected for inclusion into the seven sets of case studies by considering, as our previous methodology outlines, the technical merits of each finish. Figure 9.1, for example, shows the life cycle cost performance of four floor finishes considered technically suitable, for a heavy wear, very clean and dry space.

To derive the maximum information from the completed data set, all data were input into a computer-based model. The model, together with its own relational database, was used to carry out the iterative NPV calculations needed in producing a set of life cycle cost profiles over a series of periods of analysis up to 40 years for each case study.

In each case study, an area of 500 m^2 and a real discount rate of 4% have been used.

Intangible benefits

The next step in each case study is to consider the impact of intangible benefits. Although not measurable in any precise terms, the intangible benefits of a particular finish relative to its competitors (design, lack of disruption from longer life, and so on) might be expected to generate either increased custom or increased revenue, sufficient to offset its additional costs. In other words, at how much would the intangible benefits of a particular finish have to be valued, to justify the choice of that finish on economic grounds? This is referred to as the net benefit requirement of that finish, and may be identified by constructing net benefit requirement graphs. Assume, for example, that the decision-maker wants to know the conditions under which ceramic tiles should be installed in a particular functional space. A net benefit graph for ceramic tiles is constructed as follows:

Step 1   Calculate in the normal way the life cycle cost profiles of all the finishes being considered for that functional space.

Step 2   Identify those periods of analysis for which ceramic tiles have the lowest life cycle cost in Step 1: for such periods the net benefit requirement of ceramic tiles is, by definition, zero.

Step 3   For each period of analysis not covered by Step 2, identify the lowest cost option.

Step 4   Calculate for each period of analysis the amount by which the annual cost of ceramic tiles would have to be reduced in order to bring the life cycle cost of ceramic tiles into equality with the low-cost finish identified in Step 3. This is the net benefit requirement for that period of analysis.

Step 5   Plot the values in Step 4 in a net benefit requirement graph.

Analysing the results

Environment A

Floors
Figure 9.1 shows that of the four floor finishes considered in this example, ceramic tiles represent one of the most economical choices of finish and are always most economic for periods of analysis in excess of 40 years.

The net benefit requirement is illustrated in Fig. 9.2. This indicates, for example, that if the period of analysis is 10 years, the net benefit requirement of ceramic tiles would have to be at least $0.70/m^2$ per annum: these benefits may accrue from the advantages of ceramic tiles in terms of aesthetic qualities, ease of maintenance or length of replacement cycle. The net benefit graph can also be used to identify the period of analysis that is needed, given knowledge of the expected net benefit. For instance, given an expected net benefit of £$0.10/m^2$ per annum, ceramic tiles would become the most economic choice of finish for any period of analysis in excess of 30 years.

Walls
In the same environmental conditions, Fig. 9.3 shows that ceramic tiles represent one of the best choices of internal wall finish. Ceramic tiles in this instance rank marginally second. For a period of analysis of between 29 and 34 years, ceramic tiles are the best choice in tangible life cycle cost performance. All the finishes in this case study can be seen to rank very closely.

Figure 9.4 indicates that if the intangible benefits of ceramic tiles are evaluated in tangible terms at anything in excess of £$0.30/m^2$, ceramic tiles become the most economic finish for any project life in excess of 14 years. For building usages having a short time horizon, a benefit evaluated at £$1.00/m^2$ per annum would result in ceramic tiles being most cost-effective for any period of analysis of more than 8 years.

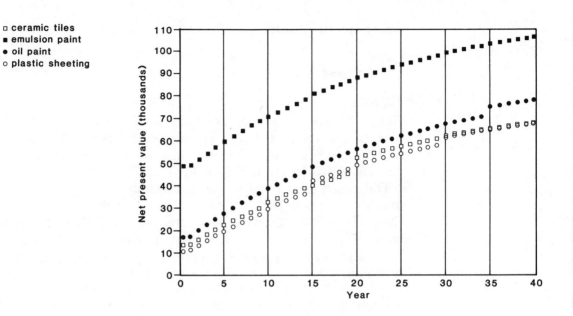

Fig. 9.1   Life cycle cost profile – environment A: floors.

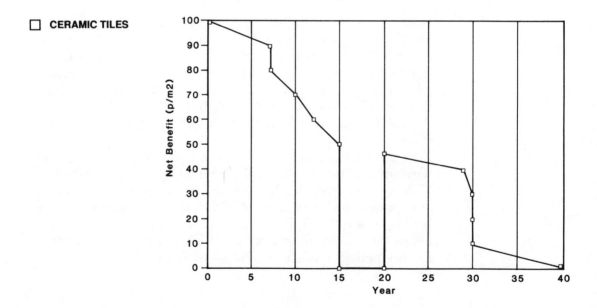

Fig. 9.2   Net benefit requirement – environment A: floors.

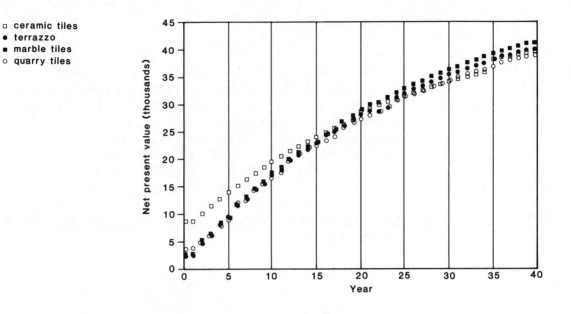

Fig. 9.3 Life cycle cost profile – environment A: walls.

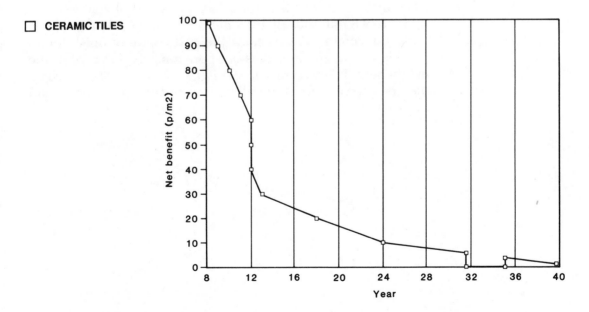

Fig. 9.4 Net benefit requirement – environment A: walls.

Environment B

Floors
Five floor finishes were considered in this example (Fig. 9.5). Both ceramic tiles and quarry tiles, in this instance, rank as the most cost-effective choice of finish. If a period of analysis of between 15 and 20 years is considered, ceramic tiles become marginally more cost-effective than any of the other finishes.

Figure 9.6 shows that a net benefit of £0.10/m^2 would change the ranking and ceramic tiles would become the most cost-effective choice of finish for periods of analysis in excess of 30 years. An increase of net benefit to £0.70/m^2 would ensure that a ceramic-tiled floor surface is most economic for all periods of analysis over 10 years.

Walls
A total of eight wall finishes were considered in this environmental category, see Fig. 9.7. Although plastic sheeting and textured finishes are ranked slightly better than ceramic tiles, the difference in total cost between the three finishes is negligible. The life cycle cost of wallpaper, by contrast, is over twice that of ceramic tiles for a 40-year period of analysis.

Figure 9.8 indicates that if a net benefit of £0.10/m^2 could be achieved, ceramic tiles become the best choice of finish for all periods of analysis in excess of approximately 40 years. If the net benefit is increased to £0.20/m^2, ceramic tiles become the first choice of finish for all periods of analysis of more than 20 years.

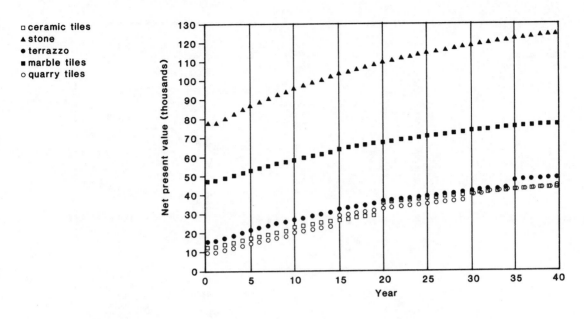

Fig. 9.5 Life cycle cost profile – environment B: floors.

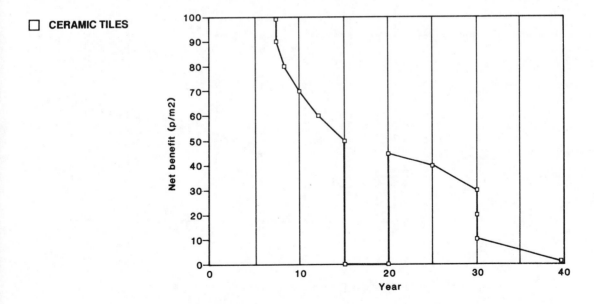

Fig. 9.6 Net benefit requirement – environment B: floors.

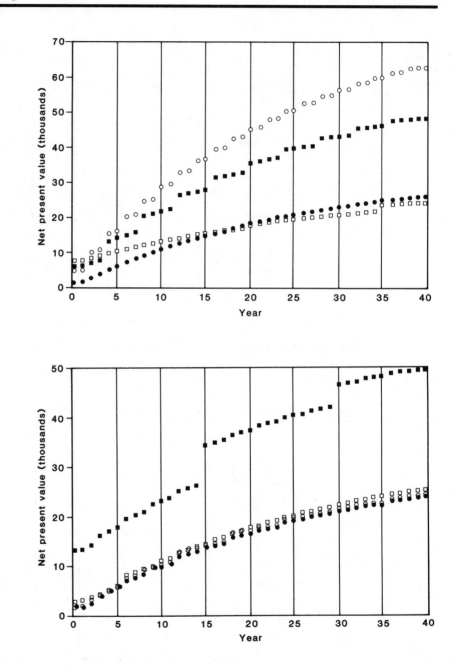

Fig. 9.7 Life cycle cost profile – environment B: walls.

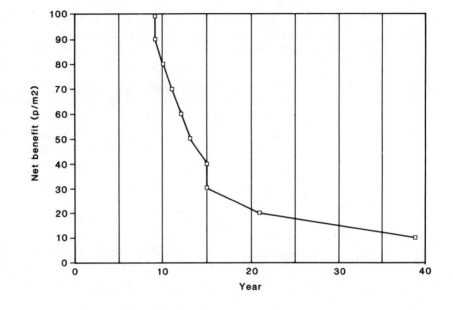

Fig. 9.8   Net benefit requirement – environment B: walls.

Environment C

Floors
Five different floor finishes are considered in this example. Figure 9.9 shows that, for a period of analysis of up to 20 years, PVC tiles rank as most cost-effective. If intangible benefits are evaluated at £0.30/m^2, ceramic tiles become the most cost-effective finish at a 40-year period of analysis (see Fig. 9.10), while if net benefits are £0.40/m^2, ceramic tiles are most economical for any period of analysis in excess of 20 years.

Note that in this case, initially the net benefits required to convert ceramic tiles into the most economic choice of finish are higher than those required in environments A and B, because the initial installation costs of PVC tiles are much lower. Note also that although the optimal replacement cycle for ceramic tiles in this case is 35 years, for a 40-year project optimal replacement would not be maintained.

Walls
In this example, Fig. 9.11 shows ceramic wall tiles ranked third with the most economical wall finish being oil paint. Figure 9.12 indicates that a net benefit of £0.30/m^2 would rank ceramic tiles first at a 40-year period of analysis.

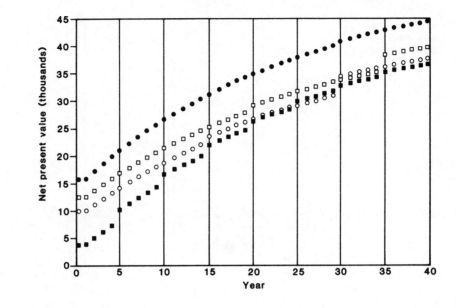

Fig. 9.9 Life cycle cost profile – environment C: floors.

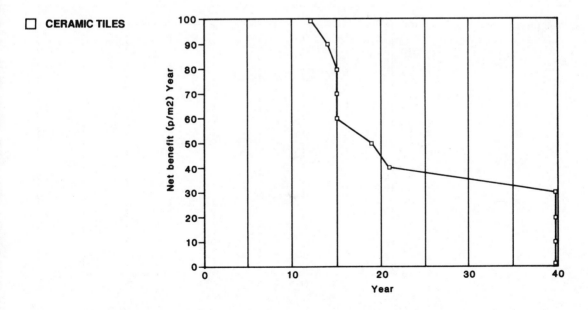

Fig. 9.10 Net benefit requirement – environment C: floors.

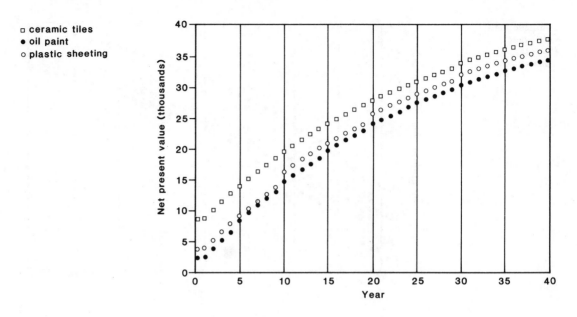

Fig. 9.11 Life cycle cost profile – environment C: walls.

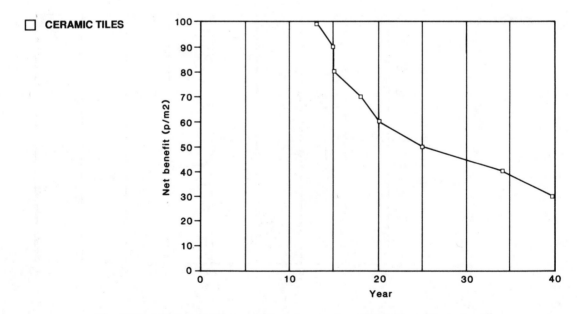

Fig. 9.12 Net benefit requirement – environment C: walls.

Environment D

Floors
In this instance, Figs 9.13(a) and 9.13(b) indicate that PVC tiles are ranked first in tangible cost terms. Figure 9.14 shows that a net benefit of £0.40/m^2 would result in ceramic tiles being the most economic choice for any period of analysis over 20 years.

Walls
Figures 9.15(a) and 9.15(b) show the life cycle cost profiles of eight wall finishes: Fig. 9.16 shows the net benefit of £0.30/m^2, that would enable ceramic tiles to become the most economic choice of wall finish at 40 years. This might be justifiable on aesthetic grounds and lack of disruption, owing to the more frequent replacement of an emulsion paint finish.

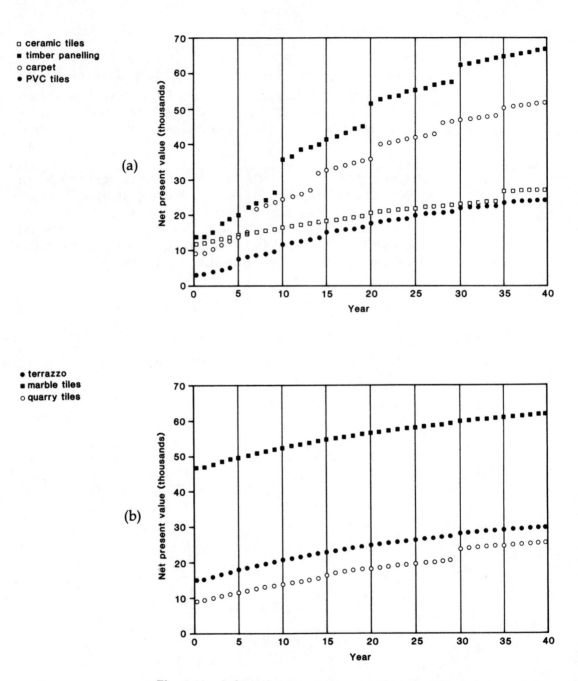

Fig. 9.13   Life cycle cost profile – environment D: floors.

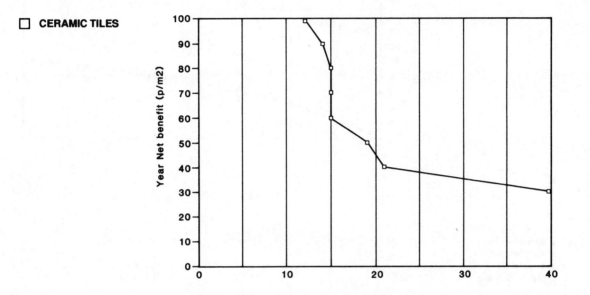

Fig. 9.14   Net benefit requirement – environment D: floors.

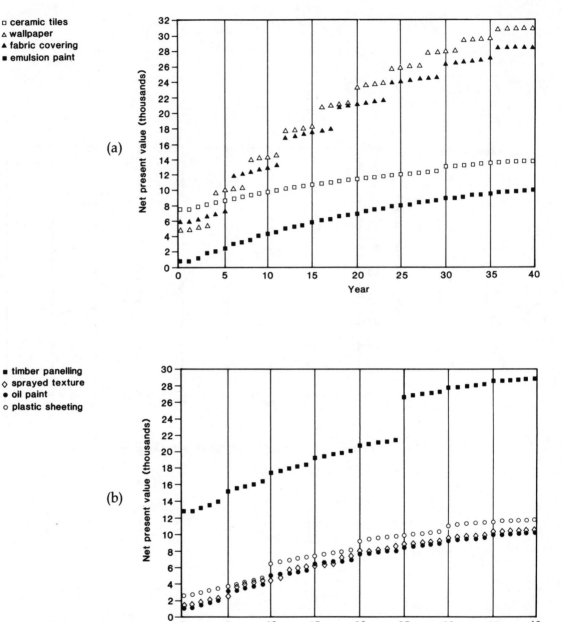

Fig. 9.15   Life cycle cost profile – environment D: walls.

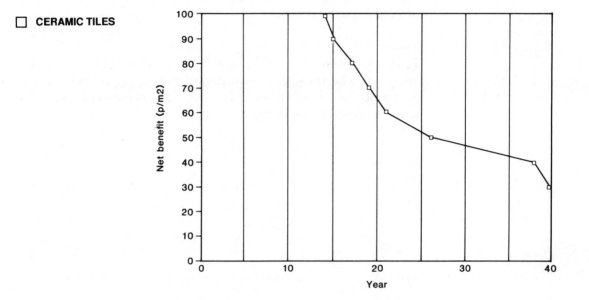

Fig. 9.16 Net benefit requirement – environment D: walls.

Environment E

Floors
The life cycle cost profiles of five floor finishes are shown in Fig. 9.17. Although PVC tiles are shown as the most economic finish, a net benefit of £0.40/m^2, as shown in Fig. 9.18, would result once again in ceramic tiles becoming the most cost-effective choice of finish.

Walls
In this example, Fig. 9.19 shows that an emulsion paint finish would be the most economical wall finish. Figure 9.20 shows that a net benefit of £0.40/m^2 is required to convert ceramic tiles into the first choice of finish.

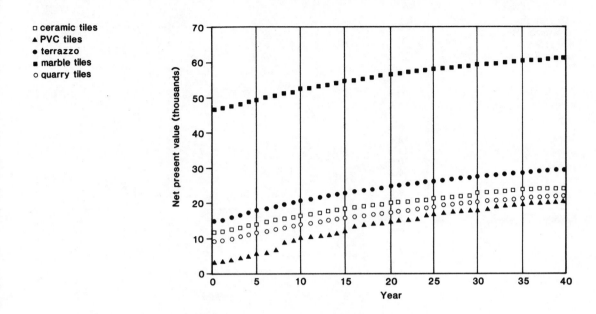

Fig. 9.17 Life cycle cost profile – environment E: floors.

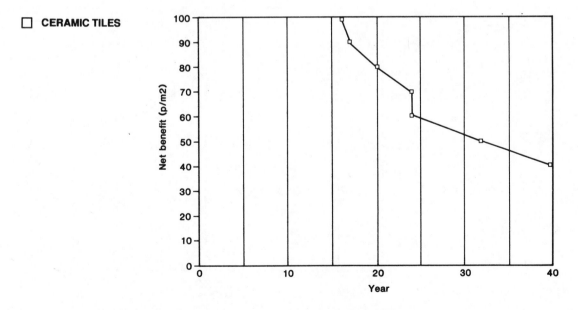

Fig. 9.18 Net benefit requirement – environment E: floors.

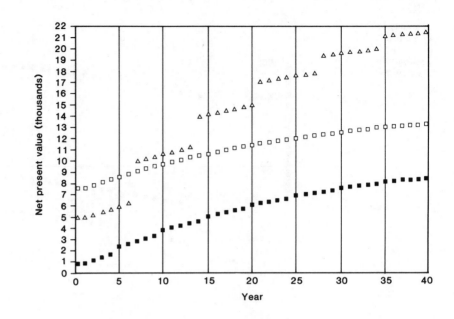

Fig. 9.19 Life cycle cost profile – environment E: walls.

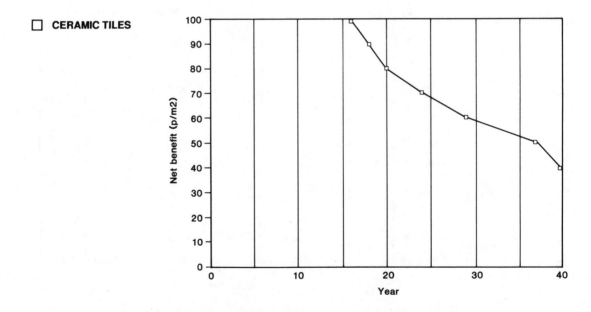

Fig. 9.20 Net benefit requirement – environment E: walls.

Environment F

Floors

The life cycle cost profiles of the four finishes considered in this example are shown in Fig. 9.21. In this instance, PVC tiles rank as the most economic finish. Figure 9.22 shows that, again because of the low initial cost of PVC tiles, a net benefit of £0.40/m^2 is required to convert ceramic tiles into the first choice at a 40-year period of analysis.

Walls

Figures 9.23(a) and 9.23(b) show that an emulsion paint wall finish, as in environment E, is the most cost-effective finish. If the intangible benefits of aesthetics and length of replacement cycle for ceramic tiles are evaluated at £0.40/m^2, Fig. 9.24 shows ceramic tiles to be the most economic choice at a 40-year period of analysis.

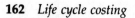
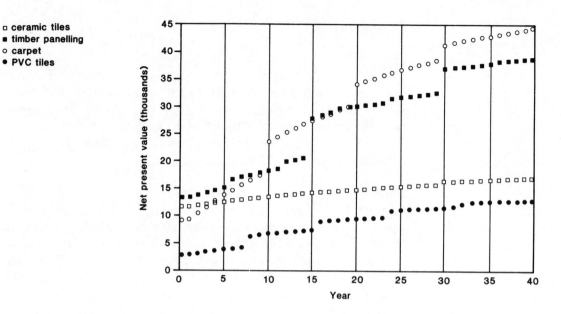

Fig. 9.21 Life cycle cost profile – environment F: floors.

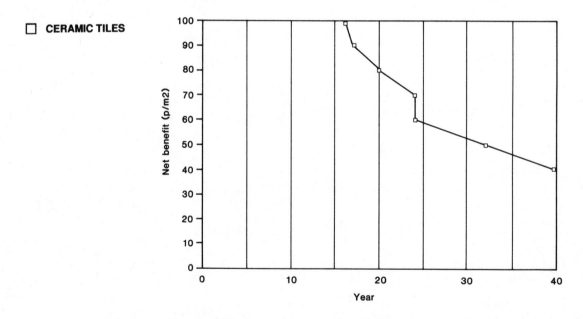

Fig. 9.22 Net benefit requirement – environment F: floors.

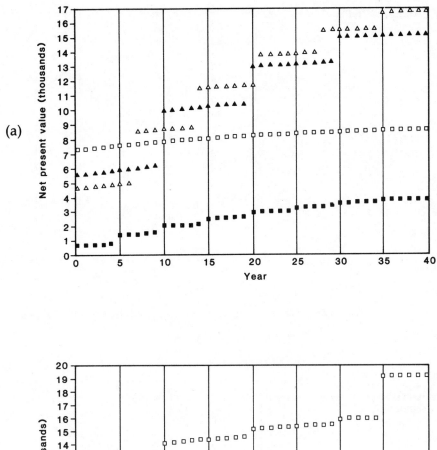

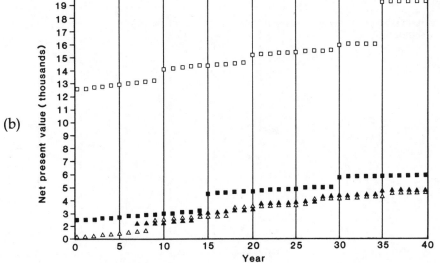

Fig. 9.23 Life cycle cost profile – environment F: walls.

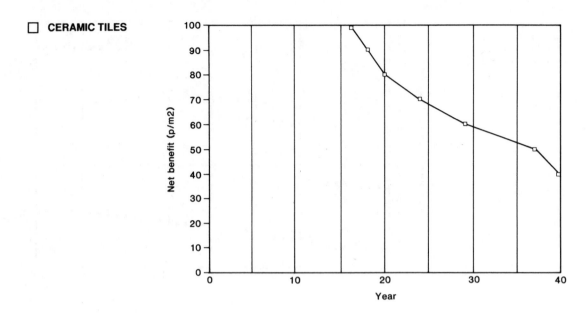

Fig. 9.24 Net benefit requirement – environment F: walls.

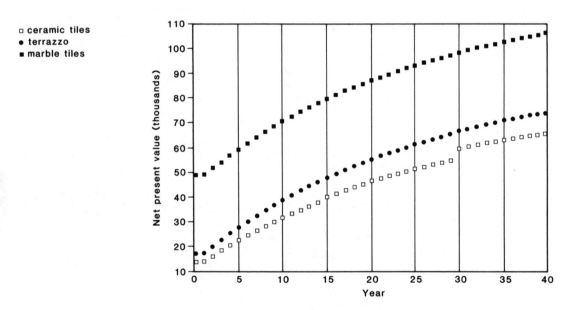

Fig. 9.25 Life cycle cost profile – environment G: floors.

Environment G

Floors

From Fig. 9.25 we see that ceramic tiles outperform all the other finishes considered in this example. Net benefits are not required, as at every point throughout the time profile ceramic tiles represent, in tangible cost terms, the most economic choice of finish.

Walls

Figure 9.26 shows that an oil-painted wall finish ranks first, followed by plastic sheeting and then ceramic tiles. However, a net benefit of £0.20/m^2, as shown in Fig. 9.27, will convert ceramic tiles into the most economic choice of wall finish at a 40-year period of analysis.

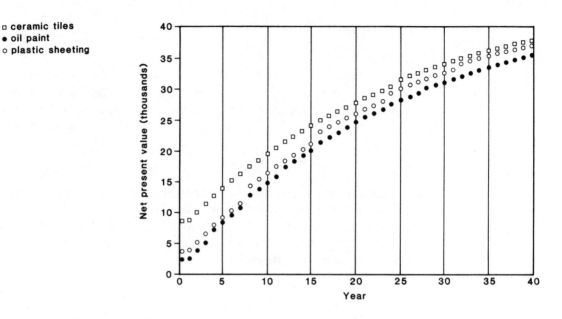

Fig. 9.26 Life cycle cost profile – environment G: walls.

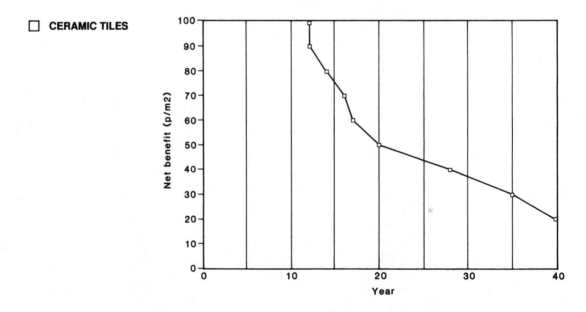

Fig. 9.27 Net benefit requirement – environment G: walls.

Chapter 10 *Conclusions*

Since 1980 there has been considerable progress in the development and application of life cycle costing techniques, but much remains to be done. There is still considerable resistance to, and scepticism about, life cycle costing techniques, among both clients and practitioners.

Many reasons for the failure to implement a total cost approach to decision-making can be identified:

- lack of appreciation of the importance of running costs;
- failure to appreciate that a balance must be struck between initial and future costs, and that a trade-off can be identified between them;
- continued lack of understanding of discounting techniques, and of how initial and future costs can be expressed in common terms;
- a feeling that the future is unknown and that basing decisions on an uncertain future is in some sense wrong.

We hope to have helped remove many of these obstacles to the implementation of the only sensible basis on which to make design decisions based on cost comparisons and to dispel the idea that there exists no simple, implementable life cycle costing system. We have attempted to add to the basic theory and give a much more extensive practical application.

Practice and methodology

The building industry is a practical industry. Theoretical advances influence the industry through their application in live situations. Any new ideas must, if they are to have a direct influence on how the industry operates, be capable of implementation. It is often incumbent upon the theoretician to provide such practical systems. In this sense, the building industry is different to a degree from many other industries and professions in which the split between theory and practice does not inhibit the process of development. This is not surprising given the nature of the industry, and it should not be a cause for complaint. Nothing tests a theorist's ideas so well as the knowledge that he is responsible for seeing them carried through to practical reality.

There is now a clear need for a practical guide to life cycle costing, rather than for further additions to the theoretical development of the subject. In that respect, this book has not been aimed at the beginner but at those who have had some exposure at least to the theory behind life cycle costing, have felt that it has some validity to the context in which they make decisions, and are now wondering how to take the next step. In this book, three new elements have been identified and developed.

An initial filter

Anyone who is now undertaking life cycle costing will know that it is a process that is relatively expensive to conduct. This is particularly true of the first few studies when computer-based systems, databases, and a general expertise have all to be developed. There is a need for an initial filtering stage that determines those options that should pass on to a full life cycle cost exercise.

This pre-selection stage should be as objective as possible. It should require the decision-maker to state why particular options should be included or omitted, preferably by relating these options in a very direct sense to the project under consideration.

Tangible and intangible costs and benefits

It has been suggested that life cycle costing is too 'clinical', presumably because it looks precise and scientific, and certainly uses some reasonably complex economic reasoning. The implication is that life cycle costing is not capable of handling more intangible and aesthetic considerations.

Nothing could be further from the truth, but equally, nothing written to date would allow the decision-maker to know this. Life cycle costing lends itself, with computer-based systems, to the 'what if' kind of questions. It naturally follows that more intangible considerations can be incorporated.

In particular, the preceding chapters have shown just how the decision-maker can address the awkward, difficult to quantify issues such as: 'I know that Option A offers me a low cost, but Option B looks better, lasts longer, and will cause much less disruption when being replaced.'

How can this dilemma be resolved? There can, of course, be no definitive answer, but the net benefit requirement technique illustrated in Chapter 9 will allow the analyst to provide the decision-maker with much more comprehensive quantitative data on which to base this sort of judgement.

Risk and uncertainty

The future is, by definition, uncertain: thus life cycle cost calculations can be estimates only. This is equally true, however, of estimated initial capital costs: the only difference is one of degree.

Nevertheless, it must be accepted that a major element in the resistance to life cycle costing is the view that it is dealing with an uncertain future. It is, therefore, necessary to grasp this particular bull by the horns and investigate, not how to avoid risk and uncertainty, but rather how to use them.

This implies the need for a sophisticated risk management system to be coupled with a practical life cycle cost system. Such risk management systems are now being developed, for example in civil engineering. Although many unresolved questions remain, these new risk management systems, which can be quite sophisticated, are easily transferred to the specific context of the building industry.

Choice of application

Finishes provide a particularly interesting example for the application of life cycle costing because:

- there is a wide choice of competing finishes
- appropriate choice is dependent on both environment and use

- the various finishes exhibit very different profiles of initial and recurrent costs, and performance characteristics
- aesthetic and other intangible considerations are of great importance.

Also, current methods of selecting wall and floor finishes for buildings are rarely based upon any method of economic evaluation other than initial capital cost comparisons with some consideration of aesthetics and performance. This does not ensure that the choices made are the best possible, in the sense that they minimise net life cycle costs during the life of a project. Our approach attempts to redress the balance.

The earlier stages of the exercise – the linking of types of facility through to detailed criteria to be met by acceptable finishes – should be particularly useful to those who have not employed systematic techniques before, or who are dealing with a type of facility with which they are not familiar. Equally, however, they are relevant to those familiar with the environment in which the finishings are to be used, as a check on the 'experience' on which judgement might otherwise be based.

Once the life cycle cost calculations have been completed, it is important that the intangible costs and benefits should be included explicitly. The cost calculations will demonstrate clearly the option that produces the lowest net measurable life cycle costs, but there may still be unquantifiable costs and benefits that the decision-maker should take into account before making the final selection, in order to make a fully-informed decision.

It may be that one finish has a lower level of life cycle costs than another, but also has a number of intangible costs. The decision-maker has to decide whether or not the intangible costs would be likely to be sufficiently large to make the finish that was apparently measurably cheaper, more expensive in practice. This is most simply expressed as: are the intangible costs likely to be greater or smaller than £x? A degree of subjectivity may be inevitable, but although life cycle costing cannot always provide all the answers, it can ensure that the right questions are asked. The existence of intangible costs and benefits means that life cycle costing should be regarded as an aid to good decision-making, not as a definitive technique that reduces decision-making to simple numerical comparisons.

The 'live' case studies indicate the relative life cycle cost performance of a number of commonly used wall and floor finishes in several different environments, including an assessment of intangible benefits. As an example, where ceramic tiles as either a wall or floor finish do not match the life cycle cost profiles of other finishes such as pvc tiles, or emulsion paint, the level of intangible benefits that should be attributed to them may result in ceramic tiles becoming the first choice of finish.

The future

The quality of decision-making arising from the use of life cycle costing is constrained by the availability of appropriate and accurate data. It has to be admitted that few comprehensive databases of the required information presently exist. However there is a plethora of information from manufacturers, trade associations, contractors and others involved in the development process. This should be examined closely and any data that would be useful in a life cycle costing exercise should be extracted and recorded.

It is worth sounding a note of optimism in this respect. Both the quality and the availability of data are significantly better than was the case even at the beginning of the 1980s. More and more practitioners have become convinced of the relevance and importance of life cycle costing and have begun to develop their own databases. The one overriding factor that must be borne in mind in any such exercise is that the data must be usable for the purposes for which they are being collected. It is of little use to collect huge quantities of data in an unsystematic manner. Rather, there needs to be a carefully constructed system to guide data collection: only then can the data be used in a constructive way, to provide estimates of costs in future projects.

One further, final positive note: the techniques suggested, and particularly life cycle costing itself, may at first sight appear to be rather difficult. However, the techniques quickly become easy to use even with very little practice, and the initial effort involved is certainly worthwhile in terms of improved decision-making. Furthermore, the application of quite straightforward computer software makes the numerical elements of the task easy. All of the applications presented in this book have been produced using custom-built software, but within a commercially available computer package. This frees the decision-maker for the more subjective, and potentially more interesting, questions relating to sensitivity analysis and the treatment of intangible costs and benefits.

References and Bibliography

American Society for Testing and Materials (1983) *Standard Practice for Measuring Life Cycle Costs of Buildings and Building Systems.* E 917–83, ASTM, Philadelphia.

American Society for Testing and Materials (1983) *Measuring Life Cycle Costs of Buildings and Building Systems.* ASTM, Philadelphia, E 917–83.

Baker, A.J. (1981) *Business Decision Making.* Croom Helm.

Baker, A.J. (1985) 'Information about risks and uncertainties under a weak ordering of preferences'. Discussion Paper No 43, University of Leicester.

Baker, A.J. (1986) 'Valuing information about risks'. Discussion Paper No 51, University of Leicester.

Bird, B. (1983) 'Costs-in-Use: Principles in the Context of Building Procurement'. *Construction Management and Economics 5.* E & FN Spon Ltd., London.

Brown, G. (1983) 'More analysis, less risk', *Chartered Surveyor Weekly*, **21**, 28 April.

Bromwich, M. (1977) *The Economics of Capital Budgeting.* Penguin Press.

Chapman, C.B. (1979) 'Large engineering project analysis'. *IEEE Trans Eng Management*, **EM-26**, pp.78–86.

Chen, P.T. and Chapman, R.E. (1980) 'Budget estimates for replacement of plant and facility equipment at the NBS campuses'. *US Department of Commerce, National Bureau of Standards*, Gaithesburg, USA. NBSIR 80–1973(R).

Cooper, D.F., MacDonald, D.H. and Chapman, C.B. (1985) 'Risk analysis of a construction cost estimate'. *Project Management*, **3**, pp.141–9.

Dell'Isola, P.E. and Kirk, S.J. (1981) *Life Cycle Costing for Design Professionals.* McGraw Hill, New York.

Flanagan, R. and Norman, G. (1982) 'Risk analysis – an extension of price prediction techniques for building work'. *Construction Papers*, **1**, pp.27–34.

Flanagan, R. and Norman, G. (1983) *Life Cycle Costing for Construction.* Surveyors Publications.

Hayes, R. (1984) 'Stemming the flow of risk'. *Risk and Loss Management,* October, pp.34–99.

Hertz, D.B. (1964) 'Risk analysis in capital investment'. *Harvard Business Review,* pp.95–106.

H M Treasury (1984) *Investment Appraisal in the Public Sector: A Technical Guide for Government Departments.* HMSO, London.

Jones Lang Wootton (1988) 'Obsolescence – the financial impact on property performance'. *Jones Lang Wootton Papers,* London.

Levy, H. and Sarnat, M. (eds) (1977) *Financial Decision Making under Uncertainty.* Academic Press, New York.

Levy, H. and Sarnat, M. (1982) *Capital Investment and Financial Decisions.* Prentice Hall, New York.

Lumby, S. (1984) *Investment Appraisal.* Van Nostrand Reinhold Co. Ltd.

Merrett, A.J. and Sykes, A.M. (1963) *The Financial Analysis of Capital Projects.* Longmans Green.

Moore, P.G. (1972) *Risk in Business Decisions.* Longman.

Perry, J.G. and Hayes, R.W. (1985a) 'Construction projects – know the risk'. *Management,* February, pp.42–5.

Perry, J.G. and Hayes, R.W. (1985b) 'Risk and its management in construction projects'. *Proc Instr CW Engrs,* **78,** pp.499–521.

Quirin, G.D. and Wiginton, J.C. (1981) *Analysing Capital Expenditures: Private and Public Perspectives,* Irwin, Illinois.

Royal Institution of Chartered Surveyors (1986) *A Guide to Life Cycle Costing for Construction.* RICS, London.

Science and Engineering Research Council, *Fundamental Approaches to Life Cycle Costing,* Grant No GR/D 59274, investigators: Norman, G. and Flanagan, R., Assistant: Robinson, G.D., end of grant date July 1987.

Wagle, B. (1966) 'A statistical analysis of risk in capital investment projects'. *Operations Research Quarterly,* **18,** pp.13–33.

Williams, B. (1988) *Premises Audits.* Bulstrode Press.

Glossary

Benefit-cost ratio   present value benefit divided by present value cost.

Bad as old (bao)   see Good as new.

Condition survey (condition audit, condition appraisal)   can be at both strategic and detailed level, the inspection of a facility, building or system and the interpretation of data to establish its condition.

Corrective maintenance   (repair) maintenance carried out after a failure has occurred and intended to restore an item to a state in which it can perform its required function, by replacement of worn or damaged parts.

Data   a representation of facts, concepts, or instructions in a formalised manner suitable for communication, interpretation or processing.

Discount rate   the discount rate is selected to reflect the investor's time value of money. The discount rate should reflect the rate of interest that makes the investor indifferent to paying or receiving a pound now or at some future time. The discount rate is used to convert future costs and revenues occurring at different times to equivalent costs at a common point in time.

Disposal costs   (salvage and residual costs and values)   the total cost or benefit to the owner of disposing of an item which is no longer required for any reason.

Emergency maintenance   maintenance which must be put in hand immediately to avoid serious consequences.

Good as new, bad as old (gan, bao)   terms used to describe a system after it has been repaired, to define whether the repair has restored it to its original condition, or whether it still contains some used or worn components. Maintenance normally restores an item to a condition which is neither gan nor bao, but models usually assume one or the other.

Life cycle cost   the total cost of ownership of an item, taking into account all the costs of acquisition, operation, maintenance, modification and disposal, for the purpose of making decisions (definition taken from BS 3811).

Life cycle cost analysis   (property occupancy cost analysis)   the analysis of the cost of buildings or systems in use.

Life cycle cost management   the management of buildings or systems in use.

Life cycle cost plan   (life cycle cost appraisal)   the uses of life cycle costing techniques at the design stage of a project.

Life cycle cost techniques   the techniques which are used in life cycle costing exercises, such as discounting future expenditure and revenues.

Nominal discount rate   (gross discount rate)   the discount rate which incorporates both the effects of general price inflation and the real earning power of money. A nominal discount rate should be used if estimates of future costs and benefits make an allowance for inflation (that is, they are in current pounds).

Period of analysis   (time study period, time horizon, investors' holding period, planning period, project life)   the period over which the investment is analysed. The period will depend upon the personal time perspective of the investor.

Planned maintenance   this may include cyclical, periodic, inter-mittent and recurrent maintenance and is the maintenance organised and carried out with forethought and control to a predetermined plan.

Present value benefit   each year's expected benefit, multiplied by its discount factor and then summed over all years of the period of analysis.

Present value cost   each year's expected cost, multiplied by its discount factor and then summed over all years of the period of analysis.

Preventive maintenance maintenance carried out at predetermined intervals, or corresponding to prescribed criteria, and intended to reduce the probability of failure.

Property database details of property owned or occupied structured as a database.

Real discount rate (net discount rate) the effects of inflation are *not* included in the discount rate (that is they are in constant pounds).

Residual value (re-sale value) the value which the residual life may have to another owner, to the benefit of the present owner.

Running costs (operations and maintenance costs, operating costs) the total costs of the operation, maintenance and modification of buildings or systems in use. These exclude the cost of staff associated with the use of the building (for example, the cost of teaching staff in a school). When staff costs are included these are often referred to as 'total occupancy costs'. The term 'occupancy cost' should be avoided unless it is precisely defined in each context.

Salvage value (scrap value) the value of the facility as salvage.

Sensitivity analysis sensitivity analysis is a test of the outcome of an appraisal based on alternative values of one or more parameters about which there is uncertainty; for example, change in the study period or the discount rate.

Unplanned maintenance maintenance carried out to no predetermined plan.

Appendix A *Choice Criteria*

Criteria Type	Characteristic	Specification *
Technical/ Performance	Abrasion resistance Acid resistance Alkali resistance Anti-static Bacterial/microbial resistance Closed pores Compressive strength Deep abrasion resistance Electricity/magnetism/radiation resistance Fire resistance Frost resistance Hardness Material allergies Moisture movement Slip resistance Sound insulation Thermal conductivity Thermal expansion Thermal shock Transverse strength Water absorption Wear resistance Weight	
Visual/ Environmental	Size Shape Colour Pattern Colour fastness Resistance to stains Size variation Flatness variation Crazing	
Initial/Running Cost	Materials Installation Repairs Maintenance annual, cyclical, periodic Cleaning Adaptation Renovation Replacement	

* This column is provided for the detailing of specific requirements, both technical, such as acceptable limits of thermal conductivity, transverse strength etc., and non-technical, such as colour required, shape of tile etc.

Index